Basic Physics for All

B. N. Kumar

University Press of America,® Inc.
Lanham · Boulder · New York · Toronto · Plymouth, UK

Table of Contents

Purpose and Audience

Basic Physics for all is a simple and straightforward book for all students taking physics and for non-physics students as well. The basic facts are presented in a concise form and convey the fundamental theories of physics. A candidate taking a high school or college examination may be expected to be familiar with the basic concepts that are outlined.

A lot of descriptive material is not included. The book can be used to supplement class teaching and to help those students who have difficulty in mastering concepts and principles.

The subject matter has been tried out with a cross-section of students at the Hillside Learning Center in Queens, New York.

Dr. B.N. Kumar, December 2008, Staten Island, NY

I. Statics

1. A force is anything which changes or tends to change the state of rest, or uniform motion in a straight line, of a body.

2. A force can be represented in magnitude and direction by a straight line bearing an arrowhead.

   The **length** of the line represents the **size** of the force. The **direction** shown by the arrowhead shows the **direction** of the force. *See fig.* 1.

   One end of the line represents the point at which the force is applied.

3. Two forces are said to be equal if, when acting on a particle in opposite directions, it remains at rest.

4. If a number of forces act on a particle and exactly balance out one another in their effects, they are said to be in **equilibrium.**

   Fig. 1 – The representation of a force by a line.

5. The **resultant** of a number of forces is the single force which will exactly replace them in its effects.

   The **equilibrant** of a number of forces is that force which is exactly equal and opposite to the resultant.

6. The **parallelogram of forces** states that if two forces acting at a point can be represented in magnitude and direction by adjacent sides of a parallelogram drawn through the point. *See* fig. 2.

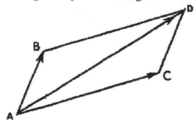

Fig. 2 – The composition of forces by the Parallelogram Law.
AD is the resultant of AB and AC.

7. (*a*) The **triangle of forces** states that if three forces acting at a point can be represented in magnitude and direction by the sides of a triangle taken in order, then the forces are in equilibrium. *See* fig. 3.

(b) The converse of the triangle of forces states that if three forces acting at a point are in equilibrium, they may be represented by the sides of a triangle taken in order.

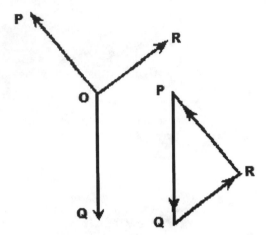

Fig. 3 – The triangle of forces.
OP, OR, and OQ are three forces
acting at O and in equilibrium.

8. Any force can be resolved into two, called its **components,** at right angles to one another. If P is a force making an angle θ with a direction X, the component of P in the direction of X is P cosθ. In the direction of Y the component is P sinθ.

9. The **units of force** are derived from the effects a force produces.
    In the C.G.S. system the practical units are the gram weight and the kilogram weight, the kilo weight being equal to 1,000 gram weight.

    In the F.P.S. system the corresponding units are the pound weight and the ton weight.

    The **absolute unit of force** in the C.G.S system is the **dyne** and in the F.P.S. system the **poundal.**

10. The **weight of a body** is the pull that the earth exerts on   the body. This
    varies slightly from place to place on the earth's surface.

11. A body may be weighed either by:
    (a) Comparing the pull of the earth on it with on some standard piece of
        matter, as in the balance.

    (b) Opposing the pull of the earth by the tension of a spring, as in the
        spring balance.

12. The **moment of a force** is in its turning effect. It is n by the product of the
    force and the perpetual distance from the pivot to the line of action of time.
    See fig. 4.

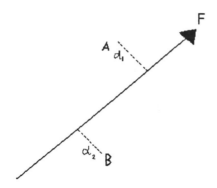

Fig. 4 - The anti-clockwise
moment of the force F about
$A = F \times d_1$ and the clockwise
moment about $B = F \times d_2$

13. The **principle of moments** states that in any system in equilibrium the sum
    of the moments in a clockwise direction about any point is equal to the sum
    of the moments in an anti-clockwise direction. *See* fig. 5.

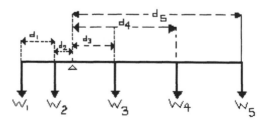

Fig 5. – The principle of moments. In equilibrium
$W_1 \times d_1 + W_2 \times d_2 = W_3 \times d_3 + W_4 \times d_4 + W_5 \times d_5$

14. The **center of gravity** of a body is the point, in or outside of the body, at which its weight may be considered to act.

15. A body is said to be in **stable** equilibrium if on suffering a **slight** displacement its center of gravity is raised. The body recovers its original position of rest when the disturbing force is removed.

      If the center of gravity is lowered by the displacement, then the equilibrium is **unstable**. The body falls away from its original position of rest.

      If the center of gravity is neither raised nor lowered, then the equilibrium is said to be **neutral**.

      A body which is tilted remains in stable equilibrium so long as the vertical from its center of gravity falls within the base. *See* fig. 6.

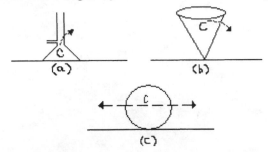

Fig. 6 – *(a)* Stable equilibrium – center of gravity raised by a slight displacement.
      (b) Unstable equilibrium – center of gravity lowered by a slight displacement.
      (c) Neutral equilibrium – center of gravity neither raised nor lowered by a slight displacement.

16. The lever, which is a simple machine, may be of one of three types:
    (a) The first system - fulcrum between load and effort. Examples: see-saw, steelyard, crowbar, chemical balance.
    (b) The second system-load between effort and fulcrum.
          Examples: wheelbarrow, trapdoor, the oar of a boat.
    (c) The third system-effort between load and fulcrum. Examples: the human forearm, steam valve, sugar-tongs (double lever). *See* fig 7.

17. The **mechanical advantage** of a machine is the ratio of the load to the effort.

$$MA = \frac{load}{effort}$$

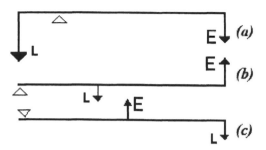

Fig. 7 – (a) First system of levers – Fulcrum between Load and Effort.
(b) Second system of levers – Load between Fulcrum and Effort.
(c) Third system of levers – Effort between Load and Fulcrum.

The **velocity ratio** is the ratio of the distance gone by the effort to the distance gone by the load.

The **efficiency of a machine** is the ratio of the work output to the work input, and is equal to the mechanical advantage divided by the velocity ratio.

N.B. – The velocity ratio is independent of all factors except the geometry of the members of the machine.

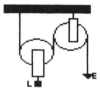

Fig. 8 – Simple fixed pulley. V.R. = 1.   Fig. 9 – Single movable pulley.
V.R. = 2.

18. A **single fixed pulley** has a velocity ratio of **one**. *See* fig. 8.

A **single movable pulley** has a velocity ratio of **two.** *See* fig. 9.

In the **first system** of pulleys, each string is fastened to the supporting beam,   and the velocity ratio is 2^n, where 'n' is the number of pulleys. *See* fig. 10.

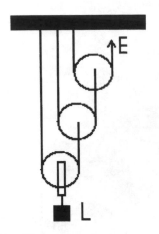
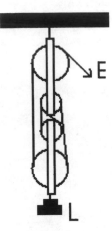

Fig. 10 – First system of pulleys. V.R. = 2^n where 'n' is
the number of pulleys.

Fig. 11 – Second system of pulleys. V.R. = number of strings
at lower block (in this case 4).

In the **second system** of pulleys the same string passes round all the pulleys, which are arranged in two sheaves. The velocity ratio is the number of strings at the lower sheaf. *See* fig. 11.

In the third system of pulleys all the strings are attached to the load and the velocity ratio is $2^n - 1$, where 'n' is the number of pulleys. *See* fig. 12.

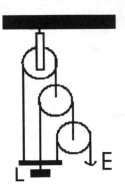

Fig. 12 – Third system of pulleys.
V.R. = 2^n-1 where 'n' *is* the
number of pulleys.

The **Weston differential pulley** has a velocity ratio of $\frac{2R}{R-r}$, where R is the radius of the larger pulley and 'r' is the radius of the smaller pulley. *See* fig. 13.

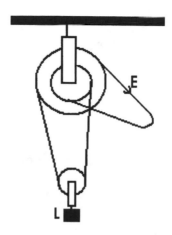

Fig. 13 – Weston differential pulley.
$$\text{V.R.} = \frac{2(Radius\ of\ larger\ pulley)}{Difference\ of\ radii.}$$

The **wheel and axle** or **windlass** has a velocity ratio $= \frac{Radius\ of\ wheel}{Radius\ of\ axle.}$
See fig. 14.

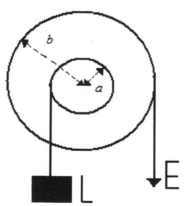

Fig. 14 – The wheel and axle.
$$\text{V.R.} = \frac{b}{a}$$

The **screw jack** has a velocity ratio of $\frac{2\pi r}{d}$, where R is the length of the arm to which the effort is applied and 'd' is the pitch of the screw. *See* fig. 15.

R

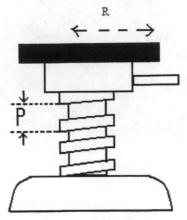

Fig. 15 – The screw jack.

$$V.R. = \frac{2\pi R}{d}$$

The **inclined plane** (*see* fig. 16) has a velocity ratio of:

(a) $\frac{Length\ of\ plane}{height\ load\ is\ raised}$ when the effort is parallel to the plane.

(b) $\frac{Length\ of\ base}{height\ load\ is\ raised}$ when the effort is parallel to the base.

The **mechanical advantage** of a lever is given by:

$$\frac{Perpendicular\ distance\ of\ effort\ from\ fulcrum}{Perpendicular\ distance\ of\ load\ from\ fulcrum}$$

Fig. 16 – Inclined plane.

   (a) Effort parallel to plane   (b) Effort parallel to base

      $V.R. = \frac{l}{h}$ $V.R. = \frac{b}{h}$

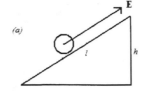
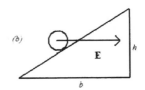

19. **The laws of statistical friction**
    (i) The frictional force always acts in such a direction as to oppose the motion of the body.

    (ii) The magnitude of the frictional force is just sufficient to prevent motion.

    (iii) The frictional force is independent of the area of the surfaces in contact.

    (iv) The frictional force depends upon the nature of the surfaces in contact.

    (v) The ratio of the frictional force to the normal reaction immediately before motion takes place is called the **coefficient of limiting friction.**

    If F is the frictional force and R is the normal reaction: $\frac{F}{R} = \mu$
    where μ is the coefficient of friction.

20.   The **angle of friction** is the angle between the frictional force F and the resultant of F and the normal reaction R. From the figure, the tangent of the angle of friction is equal to the coefficient of statical friction; i.e.: $\tan \lambda = \frac{F}{R}$
    $= \mu$ (*See* fig. 17)

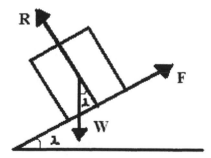

Fig. 17 – The coefficient of friction $\mu = \frac{F}{R} = \tan\lambda$ when the block is about to slip.

20. Conditions of equilibrium of parallel forces

(a) Two equal **unlike** parallel forces are called a **couple,** and such a couple has a moment equal to the product of wither force and the perpendicular distance between them.

(b) Two equal unlike parallel forces have no resultant, and can only be replaced by another couple having the same moment.

(c) If three parallel forces are to be in equilibrium, two conditions must be fulfilled:

       i.   The resolved parts of the forces in any two directions at right angles must total <u>zero</u>.

      ii.   The sum of the moments of the forces about any point in their plane must also be <u>zero</u>.

21. Conditions of equilibrium of three non-parallel coplanar forces

(a) The algebraic sum of the resolved part of the forces in any two directions at right angles must be <u>zero</u>.

(b) The algebraic sum of the moments of the forces about any point in their plane must be <u>zero</u>.

II. Dynamics

1. **Vector** quantities are those which need both a number and a direction to specify them completely. Examples are velocity, force, and magnetic field strength. **Scalar** quantities have no direction associated with them, only a number indicating size. Examples are speed, mass, and temperature.

2. The velocity of a body is its speed in a given direction. Therefore if the direction of motion changes but the speed remains the same, the velocity has changed.

3. A body is said to move with uniform velocity when it travels equal distances in the same direction in successive small equal intervals of time, no matter how small those intervals of time may be.

4. A body is said to be accelerating when its velocity is increasing, and decelerating when its velocity is decreasing.

5. A body is said to be uniformly accelerated when its velocity increases by the same amount in successive small equal intervals of time.

6. **The parallelogram law**
    Any pair of vector quantities-e.g. two velocities or two accelerations-can be compounded by the parallelogram law. They are made the adjacent sides of a parallelogram drawn to scale, and the resultant is then the diagonal of the parallelogram on the same scale through the point of intersection of the adjacent sides.

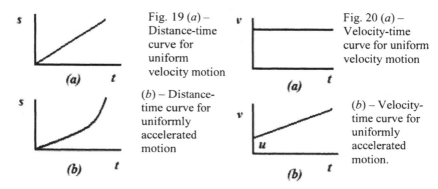

Fig. 19 (a) – Distance-time curve for uniform velocity motion

(b) – Distance-time curve for uniformly accelerated motion

Fig. 20 (a) – Velocity-time curve for uniform velocity motion

(b) – Velocity-time curve for uniformly accelerated motion.

7.   The **equations of uniformly accelerated motion.** *See* figs. 19 and 20.

Let initial velocity $= u$
Final velocity $= v$
Time taken $= t$
Acceleration $= f$
Distance gone $= s$

(a) By definition acceleration is equal to $\dfrac{Change\ of\ velocity}{Time\ taken}$

$$\therefore f = \frac{v-u}{t}$$
$$\therefore ft = v - u$$
$$\therefore v = u + ft \quad \ldots\ldots\ldots\ldots \quad (1)$$

(b) Distance gone = average velocity × time

$$\therefore s = \frac{u+v}{2} \times t$$
$$\therefore \frac{2s}{t} = u + v \quad \ldots\ldots\ldots\ldots\ldots \quad (2)$$

From (1) $ft = v - u$ . . . . (3)
Eliminating v from (2) and (3)

$$\frac{2s}{t} - ft = 2u$$
$$\therefore s = ut + \frac{1}{2}ft^2 \quad \ldots\ldots\ldots\ldots \quad (4)$$

(c) Multiplying (2) by (3) $v^2 - u^2 = 2fs$

$$\therefore v^2 = u^2 + 2fs \quad \ldots\ldots\ldots\ldots \quad (5)$$

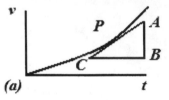
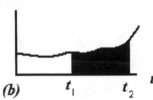

Fig. 21 – (*a*) Non-uniformly accelerated motion.   (*b*) Non-uniformly accelerated motion.
Acceleration at $P = \dfrac{Change\ in\ velocity}{time\ taken} = \dfrac{AB}{CB}$ Distance gone between times t_1 and t_2
is represented by the shaded area.

8.   The commonly used units of velocity are miles per hour, feet per second, and kilometers per hour and centimeters per second.
     The units of acceleration are feet per second and centimeters per second per second.

9.   The **momentum** of a body is the product of its mass and velocity. The unit of momentum when the mass is in grams

and the velocity is in centimeters per second is the dyne second, and when the mass is in pounds and the velocity in feet per second the poundal second.

10. The **principle of conservation of momentum** states that, provided no external forces act, the total momentum of a system of bodies remains constant.

11. **Newton's laws of motion**

    1.   Every body continues in its state of rest or uniform motion in a straight line unless acted on by an external force.

    2.   Rate of change of momentum is proportional to the impressed force.

    3.   Action and reaction are equal and opposite.

Law (1) is sometimes called the **Law of Inertia.** Law (2) is the basis on which the absolute unit of force is defined.

Since momentum is mass × velocity (mv), rate of change of momentum is $\frac{mass \times change\ of\ velocity}{time\ taken\ in\ change}$, i.e. $\frac{m(v_2 - v_1)}{t}$ when v_2 and v_1 are final and initial velocities and 't' is time taken. But $\frac{change\ of\ velocity}{time\ taken}$ = acceleration. Hence, force is proportional to mass × acceleration.

In *absolute units* Force (P) = mass (m) × acceleration (f), i.e. $P = mf$.

12. The **absolute unit of force** in the C.G.S. system is the dyne-the force required to produce in a mass of 1 gram of matter an acceleration of 1 cm/sec^2. The corresponding unit in the F.P.S. system is the poundal, which is the force required to give an acceleration of 1 ft. /sec.^{2} to 1 lb. of matter.

13. **The relation between the mass of a body and its weight**

The mass of a body is a constant which is independent of its position on the earth's surface. The weight of a body (that is, the pull of the earth on it) varies a little with its position. The weight of a body at any one place in absolute units of force is equal to the product of its mass in practical units (grams or pounds) and the value of the acceleration due to gravity.

14. **Motion under gravity**

The acceleration produced by the pull of gravity acting on a body at any point on the earth's surface is a constant denoted by the symbol 'g'. Applying Newton's second law, a body of mass 'm' is subject to an accelerating force (i.e. its weight) equal to 'mg' absolute units.

15.   Work

The work done by a force is given by the product of the force and the distance its point of application moves **in the direction of the force.** *See* fig. 22.

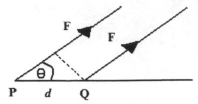

Fig. 22 – Work done = force × distance gone in the
direction of the force, i.e. F × d cos θ.

16. Units of work

The **F.P.S. absolute** unit of work is the **foot-poundal.**
The **F.P.S. practical** units of work are the **foot-pound** and **foot-ton.**
The C.G.S. absolute unit of work is the dyne-centimeter or **erg.**
The **C.G.S. practical** unit of work is the **joule.**
There are 32 foot-poundals in 1 foot-pound and 10 million (10^7) ergs in 1 joule.

17. **Power** is the rate of doing work.

The F.P.S. absolute unit of power is the foot-poundal per second and the practical unit is the **horse-power.** This is a rate of working of 33,000 foot-pounds per minute or 550 foot-pounds per second.
The C.G.S. absolute unit of power is the erg per second and the practical unit is the **watt**, which is a rate of working of 1 joule per second. The kilowatt (1,000 watts) is also a practical unit of power.

18. **Work and Energy**

The terms work and energy indicate the same physical quantity and therefore have the same units. 'Energy' is used in all branches of physics, but 'work' is usually reserved for mechanical cases of energy.
Energy is the capacity of a body for doing work.

19.   Potential and Kinetic Energy

The energy which a body possesses by virtue of its position is called its **potential energy**. Examples of potential energy are the energy of a coiled spring and the energy of a stretched elastic catapult. The energy which a body possesses by virtue of its motion is called its **kinetic energy**. It is measured by the product of one half its mass and the square of its velocity. If the mass and velocity are measured in ordinary practical units, then the kinetic energy is in absolute units, e.g. foot-poundals or ergs.

20. The conservation of energy

'Energy can be neither created nor destroyed.'

Although energy cannot be destroyed, it can be **transformed** into another kind, e.g. from mechanical energy into heat.

21. Conservation of gravitational potential energy into kinetic energy. *See fig. 23.*

A body of mass 'm' falling freely through a distance 'h' under the action of gravity loses potential energy equal to 'mgh' absolute units. It gains a velocity 'v' and therefore kinetic energy $\frac{1}{2}mv^2$:

$$\therefore\ mgh = \frac{1}{2}mv^2$$
$$\text{or}\quad v^2 = 2gh$$

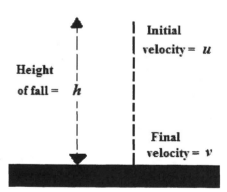

Height of fall = h

Initial velocity = u

Final velocity = v

Fig. 23 – Conservation of potential energy of a freely falling body into kinetic energy.

P.E. lost = mgh. K.E. gained $= \frac{1}{2}mv^2 - \frac{1}{2}mu^2$.

By the conservation of energy $mgh =$

$$\frac{1}{2}mv^2 - \frac{1}{2}mu^2.$$

III. Hydrostatics

1. The **density** of a substance is the mass per unit volume. The units of density in the C.G.S. system are grams per cubic centimeter and in the F.P.S. system pounds per cubic foot.

2. The **specific gravity (S.G.)** of a substance is the ratio of the mass of any volume of the substance to the mass of the same volume of water. It has no units.

3. **Relation between density and S.G.** In the C.G.S. system the density in grams per cubic centimeter and the S.G. are numerically equal. In the F.P.S. system density in pounds per cubic foot is the product of the S.G. and the density of water, also in pounds per cubic foot.

4. **Measurement of density**
   (a) Regular solids, e.g. cubes of metal:
      (i) Weigh, measure dimensions and calculate volume.
      (ii) Weigh, find volume using a displacement can.
      (iii) Weigh the body in air, then in water, and use Archimedes' principle to find the volume and hence the density.
      (iv) For a body with a density of less than 1 gm./ c.c. attach to a sinker. Find the weight of the solid in air, sinker in water and solid in water.
   (b) Irregular solids (e.g. sand):
      Use density bottle.
   (c) Liquids:
      (i) Use a hydrometer.
      (ii) Find the weight of a given volume from a burette.
      (iii) Use a density bottle.
      (iv) Use Hare's apparatus to compare the liquid with water. *See* fig. 24.
      (v) Use a U-tube.
   (d) Gases
      Find the weight of a flask from which all the air has been evacuated. Fill with the gas and reweigh, hence find the mass of gas. Fill the flask with water and weigh it again, hence finding the volume.

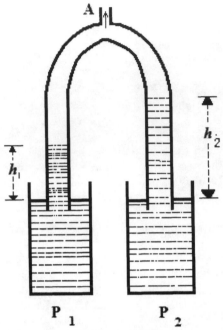

Fig. 24 – Hare's Apparatus

5. The **pressure** exerted by a fluid is the force acting on unit area. Pressure acts equally in all directions.

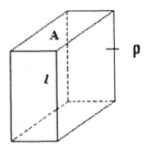

Fig. 25 – Pressure on lower horizontal
surface of a column of liquid of density

$$= \frac{Force}{Area} = \frac{Weight\ of\ column}{Base\ area\ A}.$$

$$\therefore \text{ Pressure} = \frac{Al\rho}{A} = \text{head} \times \text{density}$$

6. Pressure exerted by a column of liquid. *See* fig. 25

A column of liquid of length '*l*' and cross sectional-area A has a volume A × *l*. If its density is ρ, then the weight of the liquid column is A*l*ρ.

$$\therefore \text{Pressure at foot of column} = \frac{Force}{Area}$$

$$= \frac{Al\rho}{A} \text{ deference}$$

= Head × density.

From this it follows that at all points at the same depth beneath a liquid surface at rest the pressure is the same and proportional to the depth.

7. Units of Pressure

In the C.G.S. system pressure is measured in dynes per square centimeter and in the F.P.S. system in pounds per square inch.

Liquid and gas pressures are sometimes measured in terms of the height of the column of liquid which would produce them, e.g. feet of water or centimeters of mercury. One **foot** of water is equivalent to $\frac{12}{13.6}$ **inches** of mercury.

8. Atmospheric pressure

The pressure due to the weight of the earth's surface is called **atmospheric pressure.** It is not a constant but varies with time. Standard atmospheric pressure is that which will support a column of mercury 76 cm. long at 0° C. at sea-level. This pressure is called **one atmosphere** and is the same as a pressure of 14.7 lb/sq in. Atmospheric pressure is also expressed in millibars. One millibar is a pressure of 1,000 dynes/sq. cm. and one atmosphere is equal to 1,013.2 millibars.

9. Boyle's Law. *See* **Fig. 26.**

Provided the temperature is constant, the pressure of a given mass of gas is inversely proportional to its volume. PV = K where K is a constant.

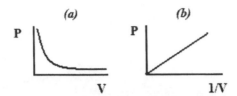

Fig. 26 – Boyle's Law. (a) Graph of pressure against volume.
(b) Graph of pressure against reciprocal of volume.

10. Devices operated by atmospheric pressure

The syringe, force pump, lift pump, siphon (*see* figs. 27, 28, 29) and vacuum brakes are all operated by the pressure of the atmosphere.

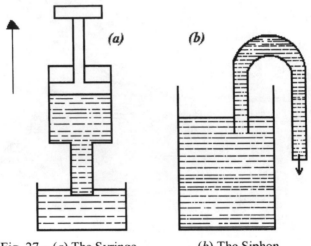

Fig. 27 – (*a*) The Syringe   (*b*) The Siphon

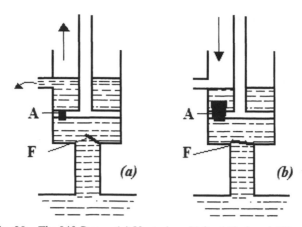

Fig. 28 – The Lift Pump. (*a*) Upstroke – Valve 'A' closed, 'F' open.
(*b*) Downstroke – Valve 'A' open, 'F' closed.

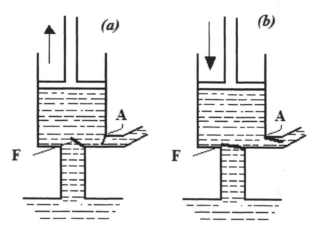

Fig. 29 – The Force Pump. (*a*) Upstroke – Valve 'A' closed, 'F' open.
(*b*) Downstroke – Valve 'A' open, 'F' closed.

11.   Barometers

Mercury barometers consist of a straight tube, closed at one end, containing mercury with a vacuum above the mercury and the lower end dipping into a mercury reservoir. The pressure is measured by the height of the mercury column. The most accurate barometer of this type is the Fortin.

Aneroid barometers consist of a sealed flat cylinder of thin metal which is evacuated. Changes in pressure cause the cylinder to expand or contract, and this movement is converted by a delicate mechanism into the rotational movement of a pointer.

12.   Archimedes' Principle

When a body is totally or partially immersed in a fluid, an up thrust comes into play equal to the weight of fluid displaced.

The principle is verified by the bucket-and-cylinder experiment.

13.   Principle of floatation

A body which floats displaces a weight of liquid equal to its own weight.

14.   The common hydrometer

$$S.G. = \frac{Length\ of\ hydrometer\ in\ water}{length\ of\ hydrometer\ in\ liquid}$$

IV. Sound

1. **The nature of sound**
   Sound is the longitudinal wave motion which arises from the vibration of a body. It requires a material medium for its propagation.

2. **Properties of sound waves**
   (a) Sound waves travel with a finite velocity.
   (b) The distance between two points in the wave which are executing the same part of their vibration identically together is called the **wavelength.**
   (c) The number of complete vibrations executed in one second is the frequency of the vibration.
   (d) The velocity 'V' of a sound wave is equal to the product of the frequency 'n' and the wavelength λ:
   $$\text{i.e. } V = n\lambda.$$

   (e) Sound waves can be reflected and approximately obey the same laws of reflection as light.
   (f) Sound waves can be refracted.

3. **The velocity of sound**
   (a) The velocity of sound in dry air at 15° C. is about 1,120 ft. /sec.
   (b) The velocity of sound in a gas is:
      (i) independent of the pressure;
      (ii) proportional to the square root of the absolute temperature.
   (c) The velocity of sound in air increases with the amount of moisture in the atmosphere.

4. **Pitch and frequency**
   The pitch of a musical note depends upon its frequency; the greater the frequency the higher the pitch.

5. **Quality**
   The quality of a musical note depends upon the number and proportion of the overtones or harmonies.

6. **Loudness**
   The loudness of a musical note is proportional to the square of the amplitude of vibration.

7. Beats

When two sources of sound of almost the same frequency are sounded together, the resulting sound fluctuates in loudness. The number of fluctuations or **beats** is equal to the difference in frequency of the two sources.

8. The Doppler effect

When a source of sound is moving towards or away from an observer, there is an apparent change in frequency. This change in frequency (which is called the Doppler Effect) depends upon the relative velocity of the source and observer and the true frequency of the source.

9. The vibration of a stretched string

The frequency of vibration of a stretched string is dependent on its length, tension and mass per unit length. If 'n' is the frequency, 'L' the length, 'm' the mass per unit length and 'T' the tension:

$$n = \frac{1}{2L} \sqrt{\frac{T}{m}}$$

V. Heat

1. The **temperature** of a body can be defined as either:
    (a) the level of heat in a body; or
    (b) the degree of hotness of the body.

2. **Thermometers** measure temperature.
    (i)   Liquid in glass thermometers use the expansion of a liquid to indicate temperature changes.
    (ii) Gas thermometers use either the expansion of gases at constant pressure or the increase in pressure at constant volume.
    (iii) Resistance thermometers use the change in the electrical resistance of a wire which accompanies a change in temperature.
    (iv) Radiation thermometers use the change in either the quality or quantity of radiation with temperature.

3. The **upper fixed point** of a mercury-in-glass thermometer is the temperature of steam above water boiling at a pressure of 76 cm. of mercury at sea-level. The **lower fixed point** is the temperature of clean melting ice.

4. **Temperature scales.** *See* fig. 30.
    The commonest temperature scales are the **Centigrade**, the **Fahrenheit**, and the **Reaumur**. On the **Centigrade** scale the upper fixed point is 100° and the lower is 0°. On the **Fahrenheit** scale the upper fixed point is 212° and the lower is 32°. On the **Reaumur** scale the upper fixed point is 80° and the lower is 0°.

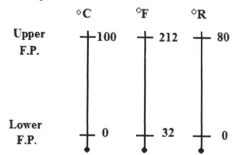

Fig. 30 – Comparison of the Centigrade, Fahrenheit, and Reaumur temperature scales.

5. Conversion of temperature scales

Since a change of 100° C. corresponds with a change of 180° F., 1° C. is equivalent to $\frac{9}{5}$° F. To convert from °C. to °F. multiply by $\frac{9}{5}$ and add 32, i.e.:

$$°F = \frac{9}{5}° \; C. + 32.$$

To convert to °C. from °F., subtract 32 and multiply by $\frac{5}{9}$, i.e.:

$$°C = \frac{5}{9} \; (°F - 32).$$

6. Mercury-in-glass thermometers

Mercury is used in thermometers because:

    (a) It has a relatively uniform expansion.
    (b) It is easily seen.
    (c) It remains liquid over a fairly wide range of temperatures.
    (d) It does not wet or stick to the glass.
    (e) It quickly takes up the temperature of the surroundings.

7. Other liquid-in-glass thermometers

Alcohol is used for low temperatures, freezing at -112° C., but it boils at 78° C.

Water is not used because:

    (a) Its expansion is not uniform (see paragraph 23).
    (b) It wets the glass.
    (c) Its range is very limited.

8. Units of heat

The **calorie** is the amount of heat necessary to increase the temperature of 1 gram of water by 1° C.

The **British Thermal Unit (B.Th.U.)** is the amount of heat necessary to increase the temperature of 1 lb. of water by 1° F.

The **Grand Calorie (or kilocalorie or 'Large' Calorie)** is equal to 1,000 ordinary calories.

The **Therm** is equal to 100,000 B.Th.U.

9. Specific heat

    (a) The specific heat of a substance is the number of calories needed to raise the temperature of 1 gram of it by 1 °C; or
    (b) The specific heat of a substance is the ratio of the amount of heat needed to raise the temperature of a given mass of a substance to the amount of heat needed to raise the temperature of the same mass of water by the same number of degrees.

10. Quantity of heat given to a body

If the mass of a body is 'm' gm., its specific heat 's' and its temperature rises by 't' °C., then the quantity of heat it receives is $m \times s \times t$ calories.

11. The fundamental equation of calorimetry

In all experiments involving an interchange of heat between bodies (as in the method of mixtures), the calculation is based on the equation:

Heat lost = Heat gained.

For accurate use of this equation the right-hand term must include any heat given to the surroundings.

12. Change of state

If heat is given to a solid which is a pure substance, its temperature rises to a fixed point called its **melting point.** If the supply of heat is not interrupted, the temperature remains constant at this value until all the solid has become a liquid. After this the temperature rises again until the **boiling-point is** reached. At the boiling-point the temperature again remains constant until all the liquid has become vapor, **always provided the pressure has not been allowed to increase.**

The heat absorbed by unit mass of a substance in changing its state at its melting- or boiling-point is called its **latent heat.**

13. Latent heat

The **latent heat of fusion** is the amount of heat needed to change unit mass of a substance from solid to liquid at its melting-point.

The **latent heat of vaporization** is the amount of heat needed to change unit mass of a substance from liquid to vapor at its boiling-point.

14. Melting- and boiling-points

The melting- and boiling-points of pure substances are constants.

Melting-points are lowered by pressure increase (the phenomenon of regelation) and the addition of impurities (e.g. salt added to water).

Boiling-points are raised by impurities and an increase in external pressure, and lowered by a decrease in external pressure.

15. The Kinetic Theory

All matter is in a state of motion. Individual molecules are in a state of violent and irregular agitation. In a solid the binding forces between molecules are sufficient to retain the solid in a definite shape. When a body is heated so that its temperature rises, the energy of agitation is increased until the binding forces become less effective and the solid loses its shape and becomes a liquid.

If the temperature increases still further, individual molecules acquire sufficient energy to leave the liquid a d form a vapor.

16. Differences between evaporation and boiling

Evaporation takes place at the surface of the liquid only. It occurs at all temperatures and is unaffected by external pressure.

Boiling takes place throughout the liquid at a definite temperature and is dependent upon the external pressure.

17. Properties of vapors

(a) Liquids exposed to the atmosphere evaporate. The rate of evaporation depends upon the nature of the liquid, its temperature and the pressure.

(b) Evaporation continues until as many molecules are leaving the liquid in one second as are returning to it in the same time. The vapor is then said to be **saturated.**

(c) A saturated vapor exerts a pressure (called the saturated vapor pressure) which varies with temperature.

(d) A liquid boils when its saturated vapor pressure is equal to the external pressure. Hence decreasing the pressure lowers the boiling-point and increasing the pressure raises it.

(e) Saturated vapors do **not** obey the gas laws; unsaturated vapors do.

18. Dalton's Law of Partial Pressures

In a mixture of gases or vapors the total pressure is the sum of the pressures which would be exerted by each vapor occupying the same space alone.

19. The expansion of solids

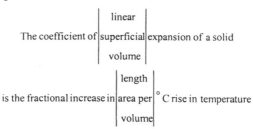

The coefficient of | linear / superficial / volume | expansion of a solid

is the fractional increase in | length / area per / volume | ° C rise in temperature

For the purpose of calculation:
Coefficient of linear expansion =

$$\frac{\text{Expansion}}{\text{Original length } x \text{ rise in temperature}}$$

20. Relationship between coefficients. *See* fig. 31

If a, β, and γ are re are respectively the coefficients of linear superficial and volume expansion respectively,

$$\text{Then } \beta = 2a$$
$$\text{and } \gamma = 3a$$

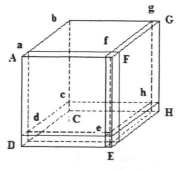

Fig. 31 – The relationship between the coefficients of linear, superficial, and volume expansion. The length *ba* expands to *b*A ; the area *afed* to AFED, and volume *abcdefgh* to A*b*CDEFGH.

21. The expansion of liquids

There are two coefficients of cubical expansion of a liquid:

(a) The coefficient of apparent expansion is the fractional increase in volume per °C rise in temperature when no allowance is made for the expansion of the containing vessel.

(b) The coefficient of absolute expansion is the fractional increase in volume per °C rise in temperature when allowance is made for the expansion of the vessel.

If the coefficient of apparent expansion is a and of real expansion γ and the coefficient of linear expansion of the material of the containing vessel is δ, then
Coefficient of absolute expansion = Coefficient of apparent expansion + coefficient of cubical expansion of containing vessel:
i.e.: $\gamma = a + \delta$

22. Relationship between liquid densities and coefficient of expansion

If the volume of a liquid increases from V_0 to V_t when its temperature is increased by $t°$ C., then, by the definition of coefficient of absolute expansion:

$$\gamma = \frac{expansion}{original\ volume \times rise\ in\ temp.}$$

$$= \frac{V_t - V_0}{V_0 t}$$

$$\therefore V_0 \gamma t = V_t - V_0$$

Or $V_t = V_0(1 + \gamma t)$.

If the densities at the lower and upper temperatures are respectively p_0 and p_1 and the mass of the liquid is 'm':

$$P_0 = \frac{m}{V_0}\ ;\ p_t = \frac{m}{V_t}$$

$$\therefore p_0 = p_t (1 + \gamma t).$$

23. The expansion of water. *See* fig. 32.

When water is cooled it contracts until 4° C. is reached. On further cooling it expands until it freezes at 0° C. At this temperature a considerable expansion occurs when water changes to ice, 1 c.c. of water becomes almost 1.1. c.c of ice. 4° C is the temperature of the maximum density of water.

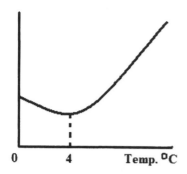

Fig. 32 – Graph showing how the volume of water varies with temperature. The minimum volume, and therefore the maximum density, occurs at 4 °C.

24. The expansion of gases

(a) When a gas is heated so that the **pressure remains constant,** it expands uniformly.

All gases expand by v h of their volume at 0° C. for each °C. rise in temperature. This is **Charles's Law.**

(b) If a gas is heated so that its **volume remains constant** but the pressure increases, it is found that the pressure increases uniformly.

The pressure of all gases increases by $\frac{1}{273}$ of their pressure at 0° C. for each °C. rise in temperature. This is the **Law of Pressures.**

(c) Combining Charles's Law with Boyle's Law (*see* III, 9) gives the General Gas Law. If the pressure, volume, and *absolute* temperature of a given mass of gas are respectively P, V, and T, then:

$$\frac{PV}{T} = \text{Constant}$$

And $\frac{P_1 T_1}{T_1} = \frac{P_2 V_2}{T_2}$

where the suffixes $_1$ and $_2$ refer to different conditions.

25. Absolute zero and the absolute scale of temperature

From Charles's Law if a gas is cooled and **remains a gas** to - 273° C., then its volume becomes zero. The temperature - 273° C. is called the **absolute zero of temperature,** and is the lowest temperature which can be measured. Temperatures measured from this zero are called **absolute temperatures (°A.)** and can be converted to °C by the equation: °C. = °A. – 273

See fig. 33.

N.B.-All gases liquefy and solidify before 0° A. is reached.

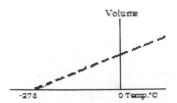

Fig. 33 – The volume-temperature graph for a gas when the pressure is kept
constant. -273° C. is called the absolute zero of temperature.

26. The mechanical equivalent of heat

Heat is a form of energy, and hence, by the conservation of energy,
cannot be destroyed, but can be trans-formed into or produced from
other forms of energy. The amount of mechanical work which must be
done to produce unit quantity of heat is called the **mechanical
equivalent of heat.** It is denoted by the symbol `J' and has the value
4.18 joules per calorie on the C.G.S. system and 778 ft.-lb.. per B.Th.U.
on the F.P.S. System.

27. Dew-point

When a mixture of air and water vapor is cooled to such a
temperature that the air becomes saturated and the water vapor exerts
its maximum pressure, some vapor begins to condense. The
temperature at which this occurs is called the **dew-point.**

28. Relative humidity

The relative humidity of the atmosphere is the ratio of the pressure
of the water vapor in the atmosphere to the saturated vapor pressure at
the same temperature. It is also the ratio of the saturated vapor pressure
at the dew-point to the saturated vapor pressure at the temperature of
the atmosphere.

29. The transmission of heat

(a) When heat is transferred from one point to another across empty
space, it is said to be transmitted by **radiation.**

(b) If heat moves from one part of a body to another without any
transfer of matter, it is said to be transmitted by **conduction.**

(c) If, in the transfer of heat from one part of a substance to another, matter is also transferred, it is said to be transferred by **convection.**

30. Conduction

All metals are good conductors of heat, the best being silver, copper and mercury. The worst conductors of heat are air, cork, rubber, ebonite, etc.

31. Convection

Convection occurs in fluids and arises because expansion occurs unevenly in heating. This causes variations in density, and the lighter parts of the fluids rise. Convection is important in climatology, and is put to practical use in the domestic hot-water system and the cooling of transformers by the circulation of oil.

32. Radiation

All materials radiate heat at a speed which depends upon the temperature. Dull black surfaces are the best radiators and polished bright surfaces the worst. A surface which is a good radiator of heat is also a good absorber and a poor radiator is a poor absorber.

VI. Light

1. **Rectilinear propagation.** *See* figs. 34, 35

    Light travels in straight lines. A point source of light produces a sharp shadow. An extended source gives rise to a shadow which is in two parts. The inner part, or **umbra,** receives no light from the source. The outer part, or **penumbra,** receives light from some parts of the source.

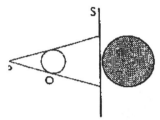

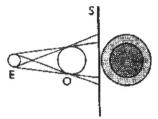

Fig. 34-The formation of a shadow by a point source of light. P-source, 0-obstacle, S-screen. The shaded area shows the size of the shadow.

Fig. 35-The formation of a shadow by an extended source of light. E-source, 0-obstacle, S-screen. The inner shaded area is the umbra and the outer shaded area the penumbra.

2. **Eclipses.** *See* figs. 36, 37

    An eclipse of the sun occurs when the moon comes between the earth and the sun in such a way that its shadow falls on the surface of the earth. A **total eclipse** occurs when the moon is at such a distance from the earth that the apparent size of the moon is bigger than the sun. When the apparent size of the moon is less than the sub, an **annular** eclipse occurs. An eclipse of the moon occurs when the earth's umbra falls on the surface of the moon.

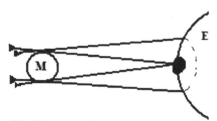

Fig. 36 – An eclipse. M- moon, E-earth. The area shown black is the region of the total eclipse and the area within the dotted region is the area of partial eclipse.

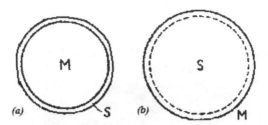

Fig. 37 – *(a)* An annular eclipse. The apparent size of the moon M is just less than the apparent size of the sun. *(b)* A total eclipse. The apparent size of the moon is greater than the apparent size of the sun.

3.   The nature of light
    Light is a wave motion which travels with a speed of 186,000 miles per second. It can travel across empty space, and is of the same nature as infra-red, ultra-violet and radio waves, and X-rays, differing only in wavelength.

4.   The laws of reflection of light
    (a) The incident ray, the reflected ray and the normal are all situated in the same plane.
    (b) The angle of incidence is equal to the angle of reflection.

5.   Real and virtual images
    A **real** image is one through which the rays actually pass, and a real image can be projected on a screen.
    A **virtual** image is one through which the rays of light only appear to pass anti cannot be projected on a screen.

6.   Properties of plane mirrors
    (i)   When an image is formed by reflection in a plane mirror, it is a virtual image.
    (ii)   Such an image is always as far behind the mirror as the object is in front.
    (iii)   If a ray of light strikes a plane mirror and the mirror is rotated about an axis, the angle turned through by the reflected ray is twice the angle turned through by the mirror.
    (iv)   Two mirrors arranged at an angle produce a number of images of an object placed between them. At 60° the number of images is five; at 90° three and at 120° two. If the mirrors are parallel, there are an infinite number of images.

7. The laws of refraction. *See* fig. 38.

(a) The incident ray, the refracted ray, and the normal are all situated in the same plane.

(b) The ratio of the sine of the angle of incidence to the sine of the angle of refraction is a constant called the refractive index (Snell's Law).

$$\mu = \frac{\sin i}{\sin r}$$

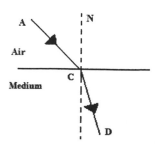

Fig. 38 – The refraction of light on passing from air into another medium. NC-normal, AC- incident ray, CD- refracted ray.

8. Properties of refractive index

(i) The refractive index depends upon the two substances in which the light is traveling; e.g. from air to glass is not the same as from water to glass.

(ii) If $a\mu g$ is the refractive index from air to glass and $g\mu a$ is the refractive index from glass to air:

$$a\mu g = \frac{1}{g\mu a}$$

(iii) For a given pair of substances the refractive index depends on the color of the light used.

9. Critical angle and total internal reflection. *See* fig. 39.

When light passes from a substance of low refractive index to one of high refractive index, it is bent towards the normal, but if the light passes from a material of high refractive index to one of low refractive index, the light is bent away from the normal. If in this case the angle of incidence is increased until the angle of emergence is 90°, then the angle of incidence is equal to the **critical angle.**

At angles of incidence greater than the critical angle, **total internal reflection** occurs and no light is refracted.

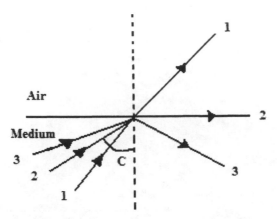

Fig. 39 – Critical angle. The ray 1 emerges from more dense to
less dense medium, ray 2 is at the critical
angle and ray 3 is totally reflected.

10. The uses of total internal reflection

Prisms are used in optical instruments in place of mirrors, the
phenomenon of total internal reflection being utilized. *See* fig. 40.

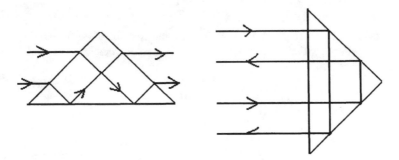

Fig. 40 –Two methods of using total internal reflection in prisms.
In each case the image is inverted.

11. The passage of light through a prism

When a ray of white light is passed through a prism:
 (i) Its path is bent, i.e. it is deviated;
 (ii) It is split up into colors, i.e. it is dispersed.

By experiment it is found that the amount of deviation depends upon
the angle of incidence and has a minimum value. For an equiangular
prism (each angle 60°) this minimum occurs when the ray passes
symmetrically through the prism.

12. Measurement of refractive index

(i) By ray tracing. For a solid block of material use pins or a ray box to trace the path of a ray of light. Measure the angles of incidence and refraction at one surface.

(ii) By tracing rays through a prism. Measure the angle of deviation for various angles of incidence and so find the angle of minimum deviation. If this angle is D and the angle of the prism is A, then

$$\mu = \frac{Sin \frac{A+D}{2}}{Sin \frac{A}{2}}$$

(iii) By the real and apparent depth method. Place a vessel of liquid over a mark on a sheet of paper and find the apparent depth by the method of no parallax. Then the refractive index is equal to the real depth divided by the apparent length.

13. Curved mirrors

If 'u' is the object distance, 'v' the image distance, 'r' the radius of curvature, and 'f' the focal length of a curved spherical mirror, then $2f = r$ and

$$\frac{1}{u} + \frac{1}{v} = \frac{1}{f}$$

if the **real-is-positive** sign convention is used.
(On this convention the focal length and radius of curvature for a convex mirror are negative and for a concave mirror positive),

14. Proof of $2f = r$ for a concave mirror. *See* fig. 41.

In fig. 41, P is the pole and PC the axis of a concave mirror whose principal focus is at F and center of curvature at C. MD represents a ray of light, parallel to the axis, striking the mirror at D and cutting the axis at F after reflection.

In the diagram, since the angle of incidence equals the angle of reflection, MDC = CDF. From the geometry of the figure, MDC = DCF.

Hence DCF = FDC and the triangle FDC is isosceles.

∴ **FD = FC**

But if D is close to P, DF = PF and therefore

PF = FC

Now, PC is the radius of curvature 'r' of the mirror and PF is the focal length 'f'.

∴ f = PC − PF

i.e. $2f = r$

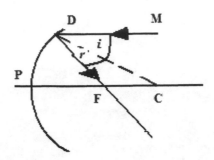

Fig. 41 – The action of a concave mirror on parallel light.
P-pole, C-center of curvature, F-focus, PC-axis.

15.   Proof of $\frac{1}{u} + \frac{1}{v} = \frac{1}{f}$ **for a concave mirror.** *See* fig. 42.

In fig. 42, OM is an object placed perpendicular to the axis PO of a
concave mirror which has P as its pole, C as its center of curvature, and
F as its principal focus. IL is the real, diminished and inverted image
which it forms. Let 'u', 'v' and 'f' be respectively the object and image
distances and the focal length.

Then MCO = ICL; LIC and MOC are right angles, and therefore the
triangles ILC and MOC are similar.

Hence $\frac{IL}{MO} = \frac{IC}{CO}$, also if A is close to P, APF and LIF are right angles;
AFP = IFL. And the triangles APF, LIF are similar.

$$\therefore \frac{IL}{AP} = \frac{IF}{PF}$$

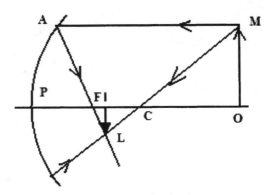

Fig. 42 – Diagram for the proof of $\frac{1}{u} + \frac{1}{v} = \frac{1}{f}$ for a concave mirror.
P-pole, C-center of curvature, F-focus, OM-object, IL-real image.

But AP = MO $\therefore \dfrac{IL}{AP} = \dfrac{IL}{MO}$, and $\dfrac{IF}{PF} = \dfrac{IC}{CO}$

$\therefore \dfrac{PI-PF}{PF} = \dfrac{PC-PI}{PO-PC}$

$\therefore \dfrac{v-f}{f} = \dfrac{2f-v}{u-2f}$

$\therefore uv - uf - 2vf + 2f^2 = 2f^2 - vf$

$\therefore uv = uf + vf$

$\therefore \dfrac{1}{u} + \dfrac{1}{v} = \dfrac{1}{f}$

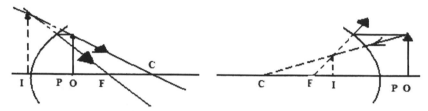

Fig. 43A – The formation of a virtual image (i) by a concave mirror. (ii) by a convex mirror.

16. Images formed in curved mirrors

	Object position	Image position	Nature of image
Concave	Between C and infinity.	Between F and C.	Real, diminished inverted.
	Between F and C.	Between C and infinity.	Real, magnified inverted.
	At C.	At C.	Real, same size, inverted.
	At F.	At infinity.	
	Between P and F.	Behind mirror.	Virtual, upright magnified.
Convex	Between P and infinity.	Behind minor.	Virtual, upright diminished.

17. Measurement of focal length

(a) Concave mirror:

(i) Focus a distant object onto a screen. The distance from the screen to the pole of the mirror is then roughly the focal length.

(ii) An illuminated object is placed in front of the mirror along its axis in such a position that it coincides in position with its own

image. The object is then at the center of curvature and hence
'r' and therefore 'f' can be found.

(iii) An illuminated object is used as in (i) above, and is placed at
varying distances from the mirror along its axis. A screen is
used to find the image position for each object distance
Image distances ('v') and object distances ('u') are measured
and substituted in

$$\frac{1}{u} + \frac{1}{v} = \frac{1}{f}$$

(b) Convex mirror:
This does not give a real image and therefore the methods of (a)
(i) and (ii) cannot be used. A plane mirror is used to form a virtual
image behind the convex mirror.

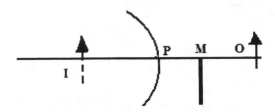

Figure 43B. – Measurement of the focal length of a convex mirror.
O-object, P-pole, M-plane mirror, I-position of *both* virtual images.

The images of a single object in the two mirrors are brought into
coincidence by the movement of the plane mirror. The distances
OM, OP (fig. 43B) are measured. For the convex mirror 'u' = OP,
'v' = OM-OP, and these values are substituted in $\frac{1}{u} + \frac{1}{v} = \frac{1}{f}$ due
account being taken of the sign according to the sign convention
used.

18. Properties of lenses
Lenses are classified according to their action on parallel light.
Those which cause it to converge are called **converging lenses** and are
thicker at the middle than at the edges. Those causing it to diverge are
diverging lenses and are thicker at the edges than at the middle.
If 'u', 'v' and 'f' respectively denote object and image distances,
$\frac{1}{u} + \frac{1}{v} = \frac{1}{f}$ on the real is positive sign convention.

19. Construction of images. *See* fig. 44.

Images of objects at finite distances can be constructed using the following rules:

    (a) A ray of light parallel to the axis is deviated through (or is apparently deviated through) the principal focus after refraction.

    (b) A ray of light striking the optical center of the lens goes through not deviated.

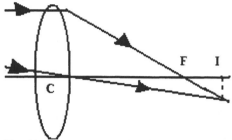

Fig. 44 – Image construction for a convex lens.

20. Proof of $\frac{1}{u} + \frac{1}{v} = \frac{1}{f}$ **for a convex lens.** *See* fig. 45.

In fig. 45, OM is the object, IL the real image and F I the focus of the convex lens PD.

In Δ's MPO, IPL, MPO = IPL (vertically opposite)

        MOP = PIL (right angles)

      $\therefore$ The triangles are similar.

Hence $\quad \frac{MO}{IL} = \frac{PO}{PI}$ . . . .(1)

In Δ's DPF, FIL, DFP = IFL (vertically opposite)

        DPF = FIL (right angles)

      $\therefore$ The triangles are similar

Hence $\frac{DP}{IL} = \frac{PF}{IF}$ . . . .(2)

But DP = MO $\therefore$ From (1) and (2):

$$\frac{PO}{PI} = \frac{PF}{IF}$$

Now PO = object distance u

      PI = image distance v

      PF = focal length f

and   IF = PI – PF = $v - f$

$$\therefore \frac{u}{v} = \frac{f}{v-f}$$

Or $uv - uf = vf$

    Dividing by uvf and rearranging:

$$\frac{1}{u} + \frac{1}{v} = \frac{1}{f}$$

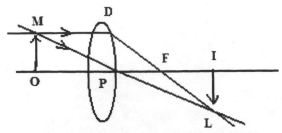

Fig. 45 – Diagram for proof of $\frac{1}{u} + \frac{1}{v} = \frac{1}{f}$ for a convex lens.

21.   Magnification. *See* figs. 45 and 42.

For either a lens or a mirror the magnification is defined by

$$\text{magnification} = \frac{image\ size}{object\ size} = \frac{IL}{OM}$$

$$= \frac{PI}{OP}$$
$$= \frac{image\ distance}{object\ distance}$$

(i)

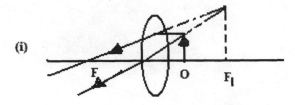

(ii)

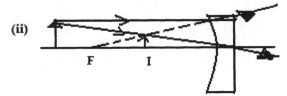

Fig. 46 – The formation of a virtual image (i) by a convex lens.
(ii) by a concave lens.

22. Images formed by lenses

Convex	Object position	Image position	Nature of image
	At infinity	At F.	Real, diminished inverted
	Between infinity and 2*f*	Between *f* and 2*f* on opposite side of lens	Real, diminished inverted
	Between 2*f* & *f*	Between 2*f* and infinity on opposite side of lens.	Real, magnified, inverted
	At F.	At infinity	
	Between F. and lens	Behind lens, between infinity and F.	virtual, magnified, upright
Concave	Any position	Between lens and F.	Virtual, diminished, upright.

23. Measurement of focal length of a lens

  (a) Convex lens:

   (i) Focus a distant object on to a screen. The distance from the screen to the lens is then roughly the focal length.

   (ii) Place a plane mirror behind the lens and perpendicular to the axis. Arrange an optical pin so that its tip lies along the axis, and move it about until it coincides with its own image. (Test by the method of no parallax.) The pin tip then lies at the principal focus.

   (iii) Arrange an illuminated object to lie along the axis of the lens and focus the image on a screen. Measure the object distance '*u*', the image distance '*v*' and apply the formula

$$\frac{1}{u} + \frac{1}{v} = \frac{1}{f}$$

Repeat for other values of the object distance and find the average value of '*f*'.

  (b) Concave lens. *See* fig. 47.

Place a plane mirror M perpendicular to the axis of the lens L and in front of it so that half the lens is covered. Place a pin 'A' behind the lens to produce a virtual image at 'A.' A second pin 'B' placed beyond the mirror will also produce a virtual image 'B.' Viewing from P, move M until the images coincide. Then the object distance is AL and the image distance (which is negative) is the difference between BM and ML. Apply the formula:

$$\frac{1}{u} + \frac{1}{v} = \frac{1}{f}$$

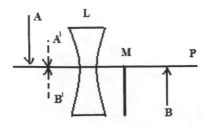

Fig. 47 – Determination of the focal length of a diverging lens.

24. Optical Instruments

(i)   See figs. 48A, 48B, 49 for details of the eye, the camera, and the projection lantern.

(ii)

Instrument	Objective	Eyepiece
Astronomical telescope	long-focus converging lens	short-focus converging lens
Galilean telescope	long-focus converging lens	short-focus diverging lens
Microscope	short-focus converging lens	short-focus converging lens

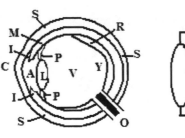

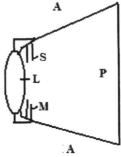

Fig. 48A – The eye. S-sclerotic, C-cornea, A-aqueous humour, L-crystalline Lens, I-iris, P-supporting ligaments of lens, Y-yellow spot, R-retina, M-choroid, O-optic Nerve, V-vitreous humour.

Fig. 48B – The camera. L-converging lens, M-diaphragm, S-shutter, AA-light-proof box, P-sensitive plate or film.

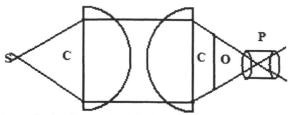

Fig. 49 – The projection lantern. S-white-light source, CC-condenser, O-slide, P-projection **lens.**

25. The pure spectrum. *See* fig. 50

The pure spectrum is formed using the apparatus of fig. 50. The function of each part is as follows:

(i) 'A' is the white-light source

(ii) 'B' is a slit which forms the object for converging lens 'C' at whose principal focus it is placed.

(iii) 'C' is the converging lens which produces parallel light.

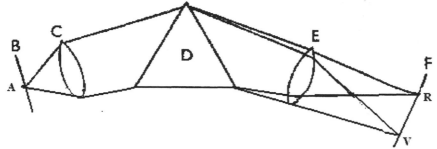

Fig. 50 – The production of a pure spectrum. A-white light source, B-slit, C-converging lens, D-prism, E-converging lens, F-screen, R-V-spectrum.

(iv) 'D' is the prism, set in the position of minimum deviation, which deviates and disperses the light.

(v) 'E' is a second converging lens which gathers the dispersed light and brings each separate color to a focus in it focal plane where the screen 'F' is placed.

26. Photometry

(a) The illuminating power (luminosity) of a source of light is the ratio of the amount of light emitted by the source to the amount emitted by a standard candle in the same time.

(b) The intensity of illumination of a surface is the amount of light energy falling on unit area in one second.

(c) The unit of illuminating power is the illuminating power of a standard candle. This is a sperm candle burning 120 grains of wax to the hour.

(d) The lumen is the amount of light falling per second on unit area perpendicular to the light and at unit distance from the source.

(e) The foot-candle is the intensity of illumination produced on a surface held one foot from a source of one candle-power.

(f) The inverse square law states that the intensity of illumination produced by a point source of light is inversely proportional to the square of the distance from it:

$$I \, a\frac{1}{d^2}$$

(g) The intensity of illumination I is proportional to the power of the source used.

$$I \propto P$$

Hence $I = k \times \dfrac{p}{d^2}$

(h) Light sources are compared by photometers, of which the simplest are (i) Bunsen's Grease Spot and (ii) Joly's Wax Block.

In each case $\dfrac{p_1}{p_2} = \dfrac{(d_1)^2}{(d_2)^2}$

27. The Electromagnetic Spectrum

Radio waves.

Short radio waves.

Ultra-short waves.

Infra-red.

Visible spectrum.

Ultra-violet.

X-rays.

γ-rays.

VII. Magnetism

1. **Chief Properties of magnets**
   (a) By contact with iron and steel, magnets can transfer some of their magnetic properties.
   (b) A bar magnet, suspended freely, will set itself north and south.
   (c) A magnet will attract iron and steel. If a bar magnet is dipped into iron filings, the regions in which they cling most strongly are called the poles. They are situated near but not at the ends of the magnet. *See* fig. 51.
   (d) The pole near the end of the bar magnet which points north is called the north-seeking or **north** pole. The end which points south is called the south-seeking or **south** pole.
   (e) If a bar magnet is broken into two, each piece becomes itself a magnet with two poles. It is impossible to produce an isolated pole.
   (f) A magnet may be made to lose its properties either by heating strongly or by rough handling or by the influence of another magnet arranged with its poles in opposition.

Fig. 51 – Iron fillings clinging to the ends of a bar magnet. PP-poles

2. **The law of poles**
   Like poles repel one another; unlike poles attract.

3. **Methods of making a magnet**
   (a) By the single-touch method.
   (b) By the double-touch method
   (c) By placing the specimen in a solenoid and passing a fairly heavy direct current.

4. **Magnetic materials**
   The chief magnetic materials are iron and steel. 'Lodestone' is a naturally magnetic iron ore consisting mostly of Fe_3O_4, an oxide of iron. Nickel and cobalt are also naturally magnetic materials, and the alloys 'Alnico' and 'Ticonal', which contain aluminum, nickel, copper and cobalt in addition to iron, are strongly magnetic.

5.   Magnetic fields

A magnetic field is a region in space over which the influence of a magnet can be detected. The earth has a magnetic field which is roughly that which would be caused by an immense bar magnet in its interior.

6.   Lines of force

The lines of force at a point in a magnetic field indicate the strength and direction of the field at that point.

If a free north pole were placed at a point in a magnetic field, then the line of force at that point gives the direction in which it would travel.

Lines of force can be plotted either by the use of iron filings or a plotting compass.

7.   Magnetic induction.

If a piece of iron or steel is brought near to, but not touching, a bar magnet, it is found to become magnetized itself. The end nearer the North Pole becomes a south pole. The specimen is said to be an **induced** magnet. Its magnetism disappears when the bar magnet is removed.

8.   Law of force between magnetic poles

If two magnetic poles have strengths m_1 and m_2 units and are situated at a distance 'd' apart in air, the force F between them is given by the relationship:

$$F = \frac{m_1 m_2}{d_2}$$

9.   Unit magnetic pole

Unit magnetic pole is that pole which, when placed 1 cm. away from an exactly similar pole in air, repels it with a force of 1 dyne. The unit of magnetic pole strength is the **weber**. Hence in the equation:

$$F = \frac{m_1 m_2}{d_2}$$

F is in dynes when m_1 and m_2 are in webers and 'd' is in centimeters.

10.   Magnetic moment. *See* fig. 52.

The moment of a magnet is the product of its pole strength and magnetic length. If the length is $2l$ and the pole strength is 'm' webers, then the magnetic moment

M = 2ml

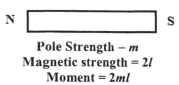

Pole Strength – m
Magnetic strength = 2l
Moment = 2ml
Fig. 52 – The meaning of magnetic moment

11. Magnetic field strength

The strength of a magnetic field is measured by the force it exerts on unit north pole. A field which exerts on a north pole of strength 1 weber a force of 1 dyne has a strength of 1 dyne per unit pole or one **oersted**.

A pole of strength 'm' webers in a field of 'H' oersted is subject to a force of 'mH' dynes.

12.   Field due to a pole strength '*m*' webers

From the equation giving the law of force between poles, the force F in dynes exerted on unit pole at a distance of 'd' cm. from a pole of strength 'm' units is

Fig. 53-A simple vibration magneto-meter. T-suspend-ing fiber, AA-draught shield, S-non-magnetic stir-rup, M-small magnet.

$$F = \frac{1 \times m}{d^2}$$

But from paragraph 11 this is also the field H at a distance 'd' cm. from a pole strength 'm'.

$$\therefore H = \frac{m}{d^2}$$

13. Magnetometers. *See* fig. 53.

A deflection magnetometer consists of a **small** bar magnet (1-2 cm. long) pivoted at the centre of a circular scale and carrying a long, light aluminum pointer at right angles. The scale is divided into four quadrants reading 0-90° each, and the whole is contained in a non-magnetic box with a mirror in its base.

The uses of the instrument are:
(a) The comparison of fields due to magnets
(b) The comparison of magnetic moments.

A deflection magnetometer is always used in such a way that the two fields at the needle are at right angles.

A vibration magnetometer consists of a small bar magnet freely suspended by an unspun silk thread. It can execute torsional vibrations about the thread as an axis. Its uses are similar to those of the deflection magnetometer.

14. Useful formula in magnetic measurements.

(a) If H_0 is the horizontal component of the earth's magnetic field and H is the magnetic field, at **right angles to H_0** produced at a point by a magnet, then the deflection θ of a deflection magnetometer at this point is given by

$$H = H_0 \tan\theta$$

(b) If a magnet has a length $2l$ cm. and a pole strength 'm' webers, then:

    a. The field it produces at a distance 'd' cm. from its mid-point along its own axis produced is

$$H = \frac{4mld}{(d^2 - 1^2)2}$$
$$= \frac{2md}{(d^2 - 1^2)2}$$

Where M is the moment of the magnet.

    b. The field it produces at a distance 'd' cm. away from its mid-point along its equator is

$$H = \frac{2ml}{(d^2 + 1^2)^{\frac{3}{2}}}$$
$$= \frac{M}{(d^2 + 1^2)}^{3/2}$$

    c. If 'l' cam be ignored compared with 'd' – i.e. if a **short** bar magnet is used- the fields become

$$H = \frac{2M}{d^3} \text{ ' end on position'}$$
$$\text{and} \quad H = \frac{M}{d^3} \text{ 'broadside on position'}$$

(c) The period of vibration 'T' of a vibration magnetometer is inversely proportional to the square root of the field in which it was used.
    If T_1 and T_2 are the periods in fields H_1 and H_2
$$(\frac{T_i}{T_2}) = (\frac{H_i}{H_2})$$

15. Neutral points

A neutral point is a point at which two magnetic fields are exactly equal and opposite, and therefore at a neutral point the total field is zero. Neutral points commonly occur when bar magnets are used in conjunction with the earth's field.

16. The earth's field. *See* fig. 54.

(a) A freely suspended compass needle sets itself **magnetic** north and south under the influence of the earth's magnetic field.

(b) The magnetic meridian at a point on the earth's surface is the vertical plane which contains the magnetic north-south line.

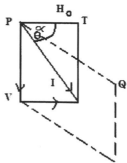

Fig. 54 – The magnetic elements. H_0- the horizontal component of the earth's field, V-the vertical component, I- the total intensity, θ- the angle of dip, ∝- the angle of declination, PQ-direction of Geographical North.

(c) The **angle of declination** at a point is the angle between the geographical north-south line and the magnetic meridian at that point.

(d) A uniform bar magnet, suspended by a torsionless thread at its midpoint, sets itself north-south and with its north pole dipping downwards at an angle of about 67° to the horizontal. The angle between the axis of the magnet and the horizontal is called the **angle of dip.**

(e) If the total intensity of the earth's magnetic field at a point is I, its horizontal component Ha and its vertical component V, then from diagram

$$\frac{V}{H_0} = \tan \theta$$

Where θ is the angle of dip and
$$I = \sqrt{V^2 + H_0 2}$$

VIII. Electrostatics

1. **Meaning of the term electrostatics**

    Modern theories of electricity state that electric charges are an essential part *of* matter. These charges can be of two kinds-positive and negative. The study of electric charges at rest is called **electrostatics.**

2. **Production of static electric charges**

    If a glass rod is rubbed with silk, the glass becomes **positively** charged and the silk **negatively charged.**

    If a rod of ebonite is rubbed with fur the rod becomes **negatively** charged and the fur **positively charged.**

3. **Conductors and insulators**

    Substances which allow the passage of electric charges through them are called conductors, and substances which do not allow charges to pass are called **insulators.** All metals are conductors, the best being silver, copper, gold, and aluminum. The best insulators are paraffin wax, mica, sulphur, ebonite, porcelain, and rubber.

4. **Interaction of charges**

    (a) Like charges repel one another; unlike charges attract.

    (b) The force between two electric charges q_1 and q_2 is proportional to their product q_1q_2 and inversely proportional to the square *of* their distance apart 'd'. If q_1 and q_2 are measured in absolute C.G.S. units, 'd' in centimeters and 'F' in dynes:
    $$F = \frac{q_1 q_2}{d^2}$$

5. **Potential**

    The potential at a point is the **work done** in bringing unit positive charge from infinity to that point.

    The potential of the earth is zero. Positive charges always flow from a place at a high potential to a place at a low potential, and negative charges in the reverse direction.

    All points on the surface of a conducting body are at the same potential.

6. **Distribution of electric charges**

    (a) Charges always reside on the **outside** of a hollow conductor.

    (b) If the surface of a conducting body is not a plane, then the charges are not evenly distributed. There are more charges per square centimeter of surface in regions which are sharply curved, the sharper the curve the greater the charge density. *See* fig 55.

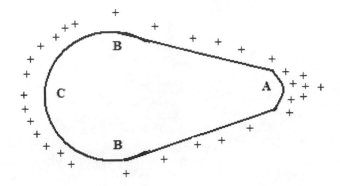

Fig. 55 – The distribution of charges on the curved surface of a conductor.

7.   Electrostatic induction. *See* fig. 56.

A previously uncharged conductor can be given a charge by the process of induction **without contact** with a charged body.

In the diagram, **A** represents the initially uncharged conductor and **B** the charged one.

(a) **B** is brought close to **A** and charges are induced on **A**.

(b) When **B** is removed, the charges disappear from **A**.

(c) **B** is brought up again, and whilst near to **A**, **A** is earthed momentarily.

(d) When the earth has been disconnected, **B** is removed. On testing, **A** is found to have a charge **equal** to that of **B** and **opposite** in sign.

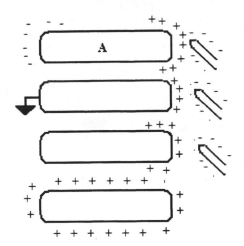

Fig. 56- Charging by induction

8. The electrostatic field and lines of force

The region in space over which a charge or a collection of charges can exert a force on other charges is called an electrostatic or electric field. Like a magnetic field, it has lines of force. These have the following properties:

(a) They begin on positive and end on negative charges.
(b) They can pass through insulating materials but not conductors.
(c) Lines of force are always perpendicular to the surface of a conductor.
(d) Positive charges always try to move down lines of force, i.e. from positive to negative ends, and negative charges move up.
(e) The number of lines of force which cross a unit area perpendicular to the field indicates the strength of the field at that point.

9. The electroscope

Simple electroscopes are of two kinds:

(a) The pith-ball electroscope, of little practical value and rarely used.
(b) The gold-leaf electroscope. This is of great practical importance and can be made into a precision instrument. It can be used **either** as a detector of electric charges **or** of potential. The deflection of the leaf is proportional to the potential of the charged body to which it is connected. *See* fig. 57.

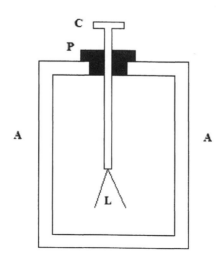

Fig. 57 – The gold-leaf electroscope.
C-cap, L-leaf, P-plug of insulating material, AA-case.

10. Capacity

The quantity of charge which must be given to a body to raise its potential by unity is called its **capacity**. If C is the capacity, Q the charge, and V the potential produced by this charge, by definition

$$C= \frac{Q}{V}$$

The unit of charge is the **farad**, which is the capacity of a conductor which needs 1 coulomb of charge to increase its potential by 1 volt. The farad is much too large for most purposes, and the **microfarad** (=1 millionth of a farad) is used instead of it.

IX. Current Electricity

1. An electric current consists of a flow of negative electric charges called **electrons.** By **convention** the current is regarded as passing in the **opposite direction** to the electron flow.

   The three main effects of an electric current are its magnetic, heating, and chemical effects.

2. **The magnetic effect of an electric current**

   Whenever an electric current flows it produces a magnetic effect, i.e. there is a magnetic field produced. The strength of this field is determined by the position in which it is measured and the size of the current flowing. Such a field has lines of force which can often be plotted using fillings or a plotting compass.

3. **Magnetic fields due to currents**

   The direction of the magnetic field produced by a current in a wire is given by **Maxwell's Corkscrew Rule:**

   'Imagine a right-handed corkscrew being turned so that its point is moving in the direction of the current, then the direction of the rotation of the handle id the direction of the line of force.'

   Figs. 58, 59, and 60 show the fields due to electric currents in a number of simple cases.

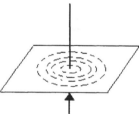

Fig. 58 – The magnetic field due to the current in a straight wire.

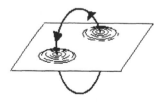

Fig. 59 – The lines of force due to the current in a single coil of wire.

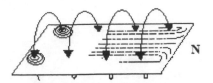

Fig. 60 – The field due to the current in a solenoid.

4. Uses of the magnetic effect of an electric current

(a) The **absolute unit** of electric current (the **abamp)** is based upon the magnetic effect of a current. When a field of one oersted is produced at the center of an arc 1 cm. long of a circle and 1 cm. radius by a current, that current is equal to one absolute unit called the **abamp.**

(b) The tangent galvanometer uses the magnetic effect of an electric current to compare the field produced by a current with the earth's field. It is therefore used in the laboratory for standardization but not normally for measuring current.

(c) The magnetic effect produced by a current flowing in a coil of wire wound round an iron core is the basic principle of all apparatus using electromagnets.

5. The electromagnet

If a soft-iron core is placed in a solenoid, the solenoid becomes, on the **passage of an electric current,** *an* **electromagnet which loses its magnetism when the** current is switched off. The polarity of the solenoid is given by the rule shown in fig. 61. The simplest pieces of apparatus using an electromagnet are the electric bell and the relay.

When the current passed through an electromagnet is sufficiently large, it is found that no further increase in current will increase the strength of the magnet. The soft-iron core is then said to be **saturated.**

Fig. 61 – The polarity of a solenoid. The arrows indicate the direction of current flow when viewed from the end.

6. The action of the electric bell. *See* fig. 62.

On passing a current through the coils AA of the electromagnet, the armature B is attracted towards the pole faces DD. The hammer E strikes the gong F and the contact C is pulled away from the point P. This stops the current and AA ceases to be a magnet. The armature is pulled back by the pieces of spring steel S. This restores the contact between C and P and therefore the current once more flows and the process is repeated.

In a single-stroke bell the contact C is short-circuited by the wire GC (shown dotted in the figure) and AA remains an electromagnet so long as the external circuit is complete.

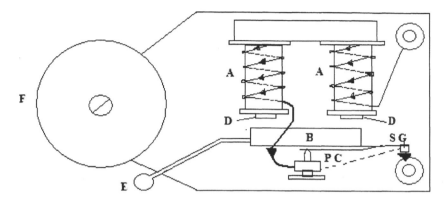

Fig. 62 – The electric bell.

7. The tangent galvanometer

The tangent galvanometer consists of a circular coil of wire held vertically and with a magnetometer box placed horizontally so that the needle is at the centre of the coil. The field produced at the centre of the coil by the current is balanced against the horizontal component of the earth's field. The galvanometer is set up as follows:

(i)   The instrument is leveled.

(ii) The coil is set to lie in the magnetic meridian. This ensures that the field due to the current and the earth's field are at right angles.

(iii) The magnetometer box is adjusted so that, with no current flowing, the pointer rests over the 0°-0° marks on the circular scale. The magnet itself then lies in the same plane as the coil.

In using the galvanometer the current must always be reversed and both ends of the pointer read, i.e. for each current 4 values of the deflection are obtained. It is the average of these which is used in calculations.

8. The theory of the tangent galvanometer

Since the coil is set in the magnetic meridian and the field due to the current is perpendicular to the coil, the earth's field and current field are at right angles. If the pole strength of the magnet is 'm' weber, the horizontal component of the earth's magnetic field H_0 oersted and the field due to the current H oersted, then from the force diagram:

$$\frac{mH}{mH_0} = \tan \theta$$

where θ is the deflection of the needle. $\therefore$ **H = H$_0$tanθ.**

The field at the center of a circular coil of radius 'r' cm. and 'n' turns, carrying a current of 'i' amps. is given by

$$H = \frac{2\pi n i}{10r} \text{ oersted}$$

Substituting in $H = H_0 \tan\theta$

$$\frac{2\pi n i}{10r} = H_0 \tan\theta$$

$$\text{Or} \quad i = \frac{10 r H_0}{2\pi n} \tan\theta$$

The term $\frac{10 r H_0}{2\pi n}$ is called the **reduction factor** of the galvanometer.

9. **The force on a current-carrying conductor**

When a wire is carrying a current in a magnetic field a force acts on it. The size of this force depends on the strength of the current and field. If the current and field are at right angles the direction of the force is given by **Fleming's Left-hand Rule.**

10. **Fleming's Left-hand Rule.** *See* fig. 63.

If the thumb, first, and second fingers of the left hand are held mutually perpendicular, with the first finger in the direction of the field and the second finger in the direction of the current, then the thumb gives the direction of motion.

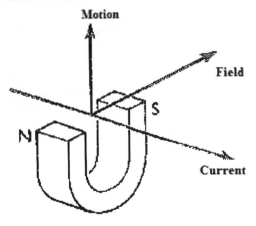

Fig. 63 – Fleming's Left-hand Rule

11. Applications of the force on a current-carrying conductor
(a) The moving-coil galvanometer. *See* figs. 64, 65

This consists essentially of a small coil of fine wire wound on an insulating frame and pivoted so that it can swing between the poles of a powerful permanent magnet. The coil rotates round a soft-iron cylinder and the pole pieces are cylindrical. The lines of force all enter the cylinder at right angles to its curved surface, and the force on the coil is always in a direction perpendicular to the current direction and the field (see fig. 65). The force caused by the passage of the current is opposed by a hairspring so that the deflection of the coil and needle is proportional to the current passing, and the needle returns to the same zero when the current is cut off.

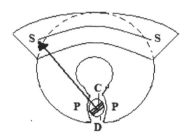

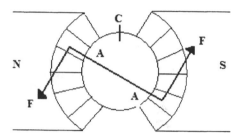

Fig. 64 – The principle of the moving-coil galvanometer. PP- cylindrical pole pieces, D-soft-iron cylinder, C-coil, S-linear scale.

Fig. 65- The forces FF always act at right angles to the field and coil as shown. The field is made radial by using cylindrical Pole pieces NS and a soft-iron cylinder C.

(b) The electric motor. *See* fig. 66.

The essential features of a simple electric motor are:
(i)   A powerful magnet which, in small motors, is usually a permanent one, but always an electromagnet in large motors.
(ii) A coil of wire which can rotate about an axis perpendicular to the field.
(iii) A commutator to which the ends of the coil are joined and through which the direction of the current through the coil can be reversed every 180° of rotation.

In fig. 66 the left-hand conductor carries the current into the plane of the paper and the right-hand conductor carries the same current out of the plane of the paper. The conductors are the sides of the coil of the motor. By Fleming's Left-hand Rule the force on the right-hand side is vertically down and on the left-hand side vertically up. These forces are equal and opposite, and constitute a couple which causes the coil to rotate anticlockwise. The momentum of the coil carries it past the vertical, and the current through CD and EF is therefore reversed by the commutator. The rotation of the coil therefore continues in the **same direction**.

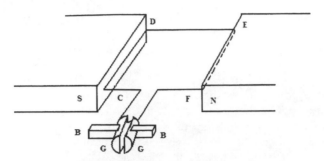

Fig. 66 – The principle of the electric motor.
NS- powerful magnet, BB- brushes,
GG- communicator, CDEF- coil.

12.   Other galvanometers. *See* figs. 67, 68.
(a)   **The moving-iron repulsion galvanometer:**
Two iron rods, one fixed and one pivoted, lie parallel to the axis of a
coil. When a current passes through the coil both become magnetized
in the same direction, and they exert on one another a force of
repulsion. This moves the pivoted rod, which has a pointer attached to
it. The amount of repulsion depends on the size of the current
passing, but the deflection is **not** proportional to the current and
therefore the scale markings are not even. When the current stops
passing, the pointer is returned to zero by gravity. The direction of the
deflection does not depend on the direction of the current.

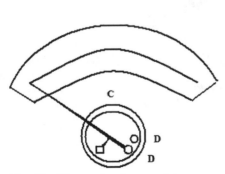

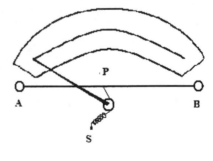

Fig. 67 – The moving-iron repulsion
Galvanometer. C- coil, DD- soft-iron
rods.

Fig. 68- Hot-wire galvanometer. AB- wire
carrying current, S-spring, P-point of
attachment of fiber to wire.

(b) **The hot-wire galvanometer:**
A fine wire is tightly stretched between two terminals. An insulating fiber, attached to the centre point of the wire, passes round a spindle carrying a pointer and is then attached to a light spring. When a current is passed through the wire it becomes heated and expands. Under the action of the spring the centre point is pulled to one side and this causes the insulating fiber to turn the spindle and pointer. In this instrument, too, the direction of rotation is independent of the direction of the current and the scale is not uniform.

13. **Ohm's Law**
Provided the temperature is constant, the current in any conductor is proportional to the potential difference between its ends. If the current I is measured in amperes, the potential difference V in volts and the resistance R in ohms, then, by Ohm's Law:

$$I = \frac{V}{R}$$

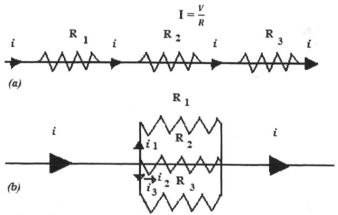

Fig. 69 – (a) Resistances in series. (b) Resistances in parallel.

14. **Applications of Ohm's Law**
(a) **Resistances in series.** *See* fig. 69 (a).
When the same current flows through two resistances they are said to be in a series. Let $r_1 r_2$ and r_3 be three resistances in series and let 'r' be the equivalent **resistance**. Then if the resistances are measured in ohms and the current 'i' amperes, by Ohm's Law:

P.D. across AB = ir_1 volts
"   "   = BC = ir_2 "
"   "   = CD = ir_3 ..
Since 'r' is the equivalent resistance, P.D. across AD is ir.
∴ $ir = ir_1 + ir_2 + ir_3$
∴ $= r = r_1 + r_2 + r_3$

(b) **Resistances in parallel.** *See* fig. 69(*b*).

If at some point in a circuit a current divides between a number of branches which later reunite, then these branches are said to be in parallel.

Let r_1, r_2 and r_3 be three resistances in parallel, each one providing a path for a current between A and B. If the main circuit current is '*i*' amps. and the currents in r_1, r_2 and r_3 are respectively i_1, i_2 and i_3 amps., the P.D. between A and B by the first path = $i_1 r_1$; by the second path = $i_2 r_2$; by the third path = $i_3 r_3$.

But this P.D. (= V) is the same for all paths.

$$\therefore i_1 r_1 = i_2 r_2 = i_3 r_3 = ir = V$$

Hence
$$i_1 = \frac{V}{r_1} \; ; i_2 = \frac{V}{r_2} \; ; i_3 = \frac{V}{r_3} \text{ and } i = \frac{V}{r}$$

And
$$i = i_1 + i_2 + i_3$$
$$\therefore \frac{1}{r} = \frac{1}{r_1} + \frac{1}{r_2} + \frac{1}{r_3}$$

(c) **Conversion of a galvanometer to an ammeter.** *See* fig. 70(*a*).

Let the current needed to give a full-scale deflection of the galvanometer be i_G amps., and let the maximum current it is required to measure be i amps. Then an alternative path is needed for the current $(i - i_G)$ amps. which cannot pass through the galvanometer. This is provided by a **shunt** resistance 'S' ohms. If the galvanometer resistance is G ohms, then the P.D. across the galvanometer by Ohm's Law is i_GG volts. P.D. across shunt = $(i - i_G)$ S. But these must be equal.

$$\therefore i_G G = (i - i_G) \times S$$

$$\therefore \text{ Shunt resistance S} = \frac{i_G}{(i - i_G)} \times G$$

$$\text{And Galvanometer current } i_G = \frac{S}{S+G} \times i$$

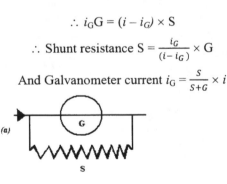

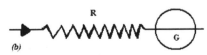

Fig. 70 – (*a*) Conversion of a galvanometer to an ammeter. (*b*) Conversion of a galvanometer to a voltmeter.

(d) **Conversion of a galvanometer to a voltmeter.** *See* fig. 70 (*b*).

If the galvanometer maximum current is i_G amps. and its resistance is G ohms, then the P.D. across its terminals is $i_G \times G$ volts. Suppose it is required to measure a voltage V. Then a *series* resistance R must be provided since V is bigger than $i_G \times G$.

Now the P.D. across $R = Ri_G$.

$$\therefore V = Ri_G + Gi_G$$
$$\text{Or } V = i_G(R + G)$$

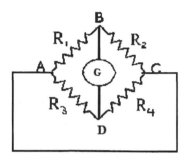

Fig. 71- The basic Wheatstone Bridge Circuit.

(e) **The Wheatstone Bridge Circuit.** *See* fig. 71.

An arrangement of four resistances, battery and galvanometer, as shown in figure 71, is called a Wheatstone Bridge Circuit. It is used to find the value of an unknown resistance R_4, the other three resistances being known. In the experiment, R_1 and R_2 are fixed and R_3 is adjusted until no current flows in the galvanometer. The bridge is then said to be **balanced.** The condition for balance is obtained as follows:

Let the current in $R_1 = i_1$; $R_2 = i_2$; $R_3 = i_3$; $R_4 = i_4$. Since at balance no current flows through the galvanometer, $i_1 = i_2$ and $i_3 = i_4$.

Now P.D. between A and B = $R_1 i_1$; P.D. between B and C = $R_2 i_1$; P.D. between D and C = $R_4 i_2$.

But, because no current flows in the galvanometer, P.D. between B and D = 0, i.e. B and D are at the same potential.

$\therefore$ P.D. between A and B = P.D. between A and D.

$$\therefore R_1 i_1 = R_3 i_2$$
$$\text{Or } \frac{i_1}{i_2} = \frac{R_3}{R_1}$$

Similarly $\frac{i_1}{i_2} = \frac{R_4}{R_2}$, so $\frac{R_1}{R_3} = \frac{R_2}{R_4}$

$$\text{Or } \frac{R_1}{R_2} = \frac{R_3}{R_4}$$

The chief Wheatstone Bridge Circuits are the **meter bridge,** where R_1 and R_2 are lengths of uniform resistance wire, and the **Post Office Box.**

15.   Resistance of a uniform wire

It can be shown that the resistance of a uniform wire is proportional to its length 'L', inversely proportional to its cross-sectional area 'A' and depends on the material of which it is made.

Hence $R \alpha L$; and $R \alpha l/A$

Or $R \alpha \dfrac{L}{A}$

i.e. $R = \dfrac{sL}{A}$

where 's' is a constant depending on the material of the wire. It is called the **specific resistance** (or resistivity) of the material of the wire. It is defined as being the resistance between opposite faces of a centimeter cube of material and its units are **ohm cm.**

16.   Units of electric current and potential difference

An electric current consists of a flow of electric charges. The practical unit of charge is the coulomb. The unit of current called the **ampere is a rate** of flow of 1 coulomb per second. The **milliampere** is one-thousandth part of 1 ampere.

The **volt** is the unit of potential difference. When 1 joule of work is done in moving 1 coulomb of charge between two points in a circuit, then the potential difference between those points is 1 **volt.**

For practical and legal purposes the definitions are as follows:

The ampere is that current which, flowing through a silver voltameter, liberates 0.001118 gm. of silver in 1 second. The **volt** is $\dfrac{1}{1.0183}$ of the potential difference between the terminals of a Weston cell on open circuit at 20° C. The international ohm is the resistance of a column of mercury of 1 sq. mm. cross-section and 106.3 cm. long at 0° C.

17.   Electromotive force. *See* fig. 72.

Everything which conducts electricity has resistance, although it may be so small in many cases as to be ignored. Batteries and cells have resistance which is often not constant and which depends upon their age and what use they have had. This resistance is called the **internal resistance.** A well-charged accumulator has an internal resistance less than $\dfrac{1}{100}$ of an ohm, whilst that of a used torch battery can be as high as 100 ohms.

Consider a battery of internal resistance '*r*' ohms connected to a circuit of resistance R ohms. If the current in the circuit is '*I*' amps., the P.D. across 'R' (and therefore across the terminals of the battery) is, by Ohm's Law, '*iR*' volts. But the same current flows through the internal resistance '*r*' and across this the potential difference is '*ir*' volts. Hence the total potential difference developed by the battery is *i (r* + R). This is called the **electromotive force** of the battery. If, in the figure, R is disconnected, no current will flow. If no current flows, *ir* = 0, and therefore the P.D. between the terminals is also the E.M.F. of the cell.

Hence, in any battery, E.M.F. = P.D. when no current flows, i.e. on open circuit.

The amount '*ir*' by which the P.D. is normally less than the E.M.F. of the battery is called the **lost volts.**

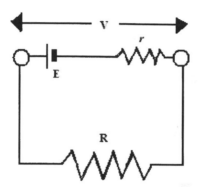

Fig. 73 – The effect of the internal resistance of a cell.

18.   The potentiometer. *See* fig. 74.

All ordinary voltmeters require some current to operate them. Hence (as shown in paragraph 17) they never indicate E.M.F., although a good volunteer takes a very small current and the error is small.

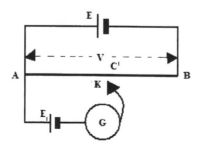

Fig. 74- The principle of the potentiometer.

The **potentiometer** is an instrument for the comparison of E.M.F.s without taking a current. It consists of a uniform resistance wire AB which is stretched over a board along which there runs a scale in centimeters. A battery is connected across the ends of the wire and sends a current through it. If this current is 'i' amps. and the resistance **per cm.** of wire is 'r' ohms, then by Ohm's Law the P.D. between the ends of any 1 cm. of wire is 'ir' volts. Across 'L' cm. of wire, therefore, the P.D. is 'irL' volts. But 'ir' is constant for all parts of the wire, and thus the P.D. across 'L' cm. of wire is proportional to 'L'.

For example:

$$\frac{P.D.across\ AC}{P.D.across\ AB} = \frac{Length\ AC}{Length\ AB}.$$

If the P.D. across AB is V volts:

$$\text{P.D. across AC} = V \times \frac{Length\ AC}{Length\ AB} \text{ volts.}$$

Suppose now a cell of E.M.F. E_1 is connected as shown in the figure, G being a galvanometer and K a sliding contact. Provided V>E_1, a point 'C' can be found at which P.D. across AC' = P.D. across AK.

Thus when K is connected at 'C' no current flows in G, K being at the same potential as C'.

Since no current flows, P.D. across AK = E.M.F. E_1 of cell.

$$\therefore \textbf{Electromotive force } \mathbf{E_1} = \mathbf{V} \times \frac{Length\ AC\prime}{Length\ AB}.$$

If E_1 is replaced by a cell E_2 and the new balance point is C^n

$$\text{Electromotive force } E_2 = V \times \frac{Length\ AC^n}{Length\ AB}$$

$$\therefore \frac{E_1}{E_2} = \frac{Length\ AC\prime}{Length\ AC^n}$$

Notes:

(a) No current flows from the cell whose E.M.F. is being found

(b) The **same pole** of each cell must be connected to A.

19. Work done in an electrical circuit

From the definition of potential difference, work done in taking unit charge from point A to point B = P.D. (V) between A and B; work done in taking 'q' units of charge from A to B = $q \times V$.

If it takes 't' seconds for this charge to pass, work done in 't' seconds = $q \times V$. Hence in 1 second charge passing = q/t and the work done per second is $\frac{q \times V}{t}$. But $\frac{Charge}{Time}$ = Current.

$\therefore$ Rate of doing work, i.e. power= current $\times$ voltage.

A current of 1 amp. passing between points whose P.D. is 1 volt is a rate of working of 1 joule per sec., i.e. 1 watt. Hence in general:

Current in amps. × P.D. in volts = Power in watts.

If the current, voltage and resistance are respectively 'I' amps., 'V' volts, and 'R' ohms:

Work done per second = $i \times$ V watts.

By Ohm's Law: $i = \dfrac{V}{R}$

∴ Work done per second = i^2 R watts

$$= \dfrac{v^2}{R} \text{ watts.}$$

20.   The heating effect of an electric current.

If a current of 'i' amps. is passed through a conductor, the potential difference between the ends being E volts, the rate at which work is done is iE joules per second.

∴ In 't' seconds work done = iEt joules.

If the number of joules required to produce 1 calorie is J (the mechanical equivalent of heat), this work will produce in 't' seconds iEt/J calories. If 'R' is the resistance of the conductor in ohms, then applying Ohm's Law:

$$\text{Heat produced} = \dfrac{iEt}{J}$$
$$= \dfrac{i^2 Rt}{J} \qquad \text{calories.}$$
$$= \dfrac{E^2 t}{RJ}$$

*This is called **Joule's Law.***

21.   Units of electrical energy

The unit in which electrical energy is bought and sold is the Board of Trade Unit. This is the energy consumed in one hour by a device working at a rate of one kilowatt, and is numerically 3,600,000 joules.

It is also called the kilowatt-hour (kwh). To calculate the amount of electrical energy consumed by a number of appliances, proceed as follows:

(a) Multiply the rating of each in kilowatts by the time in hours.

(b) Add these products and this gives the total number of units.

22.   The chemical effect of an electric current

The chemical effect of an electric current is called **electrolysis,** and is the splitting up of material (usually in a solution) by the passage of an electric current through it. This takes place in an **electrolytic cell** or **voltameter.** In the cell is the **electrolyte.** The current enters and leaves the cell by the **electrodes.** The one by which it enters is called the **anode** and the one by which it leaves is called the **cathode.**

It is at the cathode that hydrogen and the metals are deposited.

23.   Faraday's Laws of Electrolysis

(I)   The amount of a substance deposited is proportional to the quantity of charge passing, i.e. the product of the current and time.

(II)   The mass of an element deposited is proportional to its chemical equivalent.

If 'm' is the mass deposited, 'i' the current in amperes, and 't' the time in seconds:

$$\therefore m \ a \ it \text{ and } m = zit$$

where 'z' is a constant called the **electrochemical equivalent** of the element.

It is defined as the mass of element deposited by unit charge. Since the mass deposited is also proportional to the chemical equivalent, 'z' varies from element to element and is proportional to the **chemical equivalent.**

For any element:

Electrochemical equivalent =

Chemical Equivalent × Electrochemical Equivalent of Hydrogen

24.   Primary cells. *See* figs. 75, 76, 77.

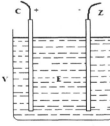
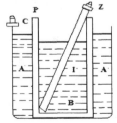

Fig. 75 – The simple cell.
V-vessel, C- +ve pole (copper plate), Z - -ve pole (zinc plate), E- diluted Sulfuric acid.

Fig. 76 – The Daniell Cell. C- +ve terminal attached to copper case, P-porous pot, Z- -ve terminal attached to zinc rod, A-copper sulfate solution, B- diluted sulfuric acid.

(a) The simple cell:
    Electrodes: +*ve* copper plate, -*ve* zinc plate.
    Electrolyte: diluted sulfuric acid.
        Faults: polarization, local action.
        E.M.F.: 0.9 volt.

Polarization is the formation of hydrogen bubbles on the plates, and is cured either by removing them mechanically or by adding a chemical-depolarizing agent to react with the hydrogen to form water.

Local action is the eating away of the zinc plate caused by the presence of impurities in the zinc. It is cured by rubbing the zinc plate with mercury. This is known as **amalgamating** the zinc.

(b) The Daniell Cell:
    Electrodes: copper outer vessel positive, amalgamated zinc rod
              negative.
    Electrolytes: inner, diluted sulfuric acid or zinc sulfate solution;
              outer, copper sulfate solution. The electrolytes are
              separated by a porous pot.
        E.M.F.: 1.1 volts.

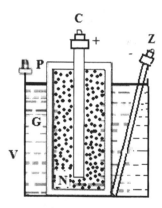

Fig. 77 – The Leclanché Cell. V-vessel, P-porous pot,
C- + *ve* terminal attached to carbon rod, Z- -*ve* terminal
attached to zinc rod, G-solution of ammonium chloride,
N-powdered carbon and manganese dioxide.

(c) The Leclanché Cell:
    Electrodes: + *ve* carbon rod, -*ve* amalgamated zinc rod.
    Electrolyte: diluted solution of ammonium chloride (sal
            ammoniac).
    E.M.F.: 1.45 volts.

The depolarizing agent is manganese dioxide, which is mixed with powdered carbon to lower the internal resistance and which is contained in a porous pot with the positive electrode.

The main virtue of the cell is that it will deliver a relatively heavy current, polarize and then recover fairly quickly. It is therefore used in applications where the current is drawn intermittently, e.g. in electric-bell circuits.

25.   Secondary Cells

A lead accumulator consists of two lead plates immersed in diluted sulfuric acid.

On charging, the positive plate is changed to lead peroxide, PbO_2, and the negative plate remains unchanged.

On discharging, lead sulfate forms at both plates.

The processes of charging and discharging are accompanied by a change in the concentration of the sulfuric acid. As the accumulator is discharged the S.G. of the acid decreases; on charging it increases, being about 1.215 when the accumulator is fully charged. The E.M.F. when fully charged is 2.15 volts. The circuit for charging is shown in fig. 78.

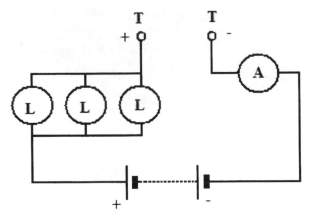

Fig. 78- The circuit for accumulator charging. B-batteries being charged, TT- D.C. supply, A-ammeter, LLL-lamp bank to control size of current.

26.   Lenz's Law

Whenever a current is induced in a circuit it is in such a direction as to oppose the motion producing it.

27. Faraday's Laws of Electromagnetic Induction
Whenever the number of lines of force through a circuit is changing, the induced E.M.F. set up is proportional to the rate at which the number of lines of force changes.

28. Fleming's Right-hand Rule. *See* fig. 79.
If the thumb, first, and second fingers of the right hand are held mutually perpendicular, with the first finger in the direction of the field and the thumb in the direction of motion, then the second finger points in in the direction of the induced current (dynamo rule).

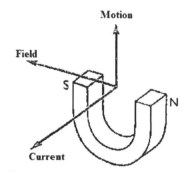

Fig.79 – Fleming's Right-hand Rule.

29. The principle of the dynamo. *See* fig. 80.
When the coil, rotating clockwise, is in the position shown, BA cuts the lines of force, and by Fleming's Right-hand Rule the current flows from B to A. Similarly in DC (which is moving upwards), the current flow is from D to C.

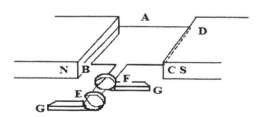

Fig. 80 – The principle of the AC generator. NS-magnet, ABCD – coil, EF- slip rings, GG-brushes.

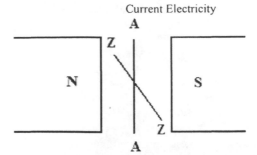

Fig. 81 – More lines of force are cut per second (and therefore the induced E.M.F.)
is greater in the position ZZ than in the position AA.

When the position of the coil is reversed, the current flow in each
conductor is reversed. The E.M.F. generated is dependent upon the rate at
which lines of force are cut. From fig. 81 it is clear that more lines of force
are cut in one second in the position ZZ than in position AA. The E.M.F. is
therefore not a constant but an alternating one which varies with time as
shown in fig. 82.

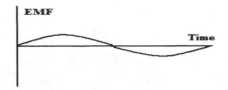

Fig. 82- E.M.F. – time curve for a simple one-turn generator with slip rings.

30.   The D.C. generator

In order to obtain an E.M.F. in one direction only, the slip rings of the
simple generator are replaced by a communicator. By this means, the
external circuit is always connected to the side of the coil which is
moving in the same direction with respect to the magnetic field.

The output of one coil of a generator of this kind is shown in fig. 83.

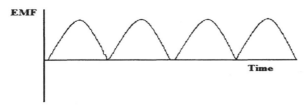

Fig. 83- Voltage-time curve for one coil of a generator fitted with a commutator.

The essential parts of a D.C. generator are:

    i.   A powerful electromagnet usually energized by the output of the machine itself.

    ii. An armature consisting of a large number of coils wound on a laminated iron core. This increases the number of lines of force cutting the coils.

    iii. A commutator of many segments, the number being determined by the number of coils.

    iv. Brushes by which the armature delivers the current.

31. Eddy Currents

By Faraday's Law an E.M.F. is generated whenever the number of lines of force through a circuit changes. If a piece of metal is placed in a changing magnetic field, the number of lines of force through the metal will change and an E.M.F. will be set up between different parts of the metal. Since the metal is a conductor, a current will flow. Currents of this kind are called eddy **currents.**

32. Applications of electromagnetic induction

(a) **The induction coil.** *See* fig. 84.

The purpose of the induction coil is to produce a high intermittent voltage from a low steady D.C. source. The main parts of an induction coil are:

    i. A laminated core of soft iron, usually a bundle of soft-iron wires.

    ii. A primary coil consisting of relatively few (200-300) turns of thick copper wire.

    iii. A secondary coil of many thousands of turns of fine copper wire.

    iv. A make-and-break mechanism like that of an electric bell.

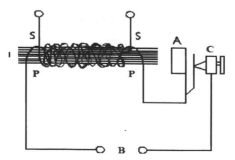

Fig. 84- The induction coil. A-armature, B-battery terminals, C-make-and-break, P-primary windings, S-secondary windings, I-soft-iron core.

The action of the make-and-break in the primary circuit is to cause a magnetic field to be set up and then destroyed in the soft-iron core. This field builds up slowly, but is quickly destroyed when the primary current ceases to flow. The sudden collapse of the lines of force causes the secondary coil to be cut, and this sets up a large induced E.M.F. across the terminals of the secondary. A condenser of waxed paper and tinfoil is often included across the contacts of the make-and-break. Its purpose is to speed up the "break action."

(b) The transformer. *See* fig. 85.

The purpose of a transformer is to provide from an alternating current source a voltage which is either greater or less than that of the source. Its essential features are:
  i.   A laminated soft-iron core round which the coils are wound.
  ii.   A primary coil
  iii. A secondary coil.

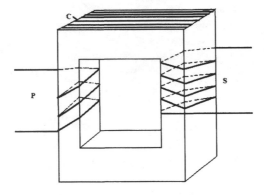

Fig. 85- The principle of the transformer. P-primary winding, S-secondary winding, C-laminated core. As shown this is a step-up transformer because there are more turns on the secondary than on the primary.

The varying magnetic field set up by the alternating current in the primary cuts the turns of the secondary coil and sets up an E.M.F. across the terminals. The relationship between the primary and secondary turns and the primary and secondary E.M.F.s is:

$$\frac{Primary\ E.M.F.}{Secondary\ E.M.F.} = \frac{Number\ of\ turns\ on\ primary}{Number\ of\ turns\ on\ secondary}$$

If the secondary E.M.F is greater than the primary, then the voltage is said **to be stepped-up.** If the secondary E.M.F. is less than the primary, the voltage is **stepped-down.** In a step-up transformer the primary winding is made of thick wire and the secondary of thin wire. The opposite is true in a step-down transformer.

33.   The electron

The electron is the fundamental unit of negative electric charge and is approximately 1.6×10^{-19} of a coulomb. It is an essential part of matter, and can be produced by a variety of methods:

(a) By frictional electrification. The charges produced remain attached to matter.

(b) By heating a metallic wire in an empty space.

(c) By applying a large voltage to two electrodes sealed into an evacuated glass tube.

(d) By allowing light to fall on certain metals (e.g. Selenium) in a vacuum.

In cases (b), *(c)* and *(d)* the electrons are produced quite free of all matter.

34.   The diode. *See* fig. 86.

The diode or two-electrode valve consists of an evacuated glass bulb into which is sealed a heater, cathode and anode. In some diodes the heater is also the cathode. An electric current is passed through the former so that the cathode becomes sufficiently hot to emit a copious supply of electrons.

If the anode is attached to the positive pole of a high-tension battery and the cathode to the negative, electrons flow from the cathode to the anode.

If the high-tension battery is reversed- i.e. the negative pole connected to the anode and the positive to the cathode- no current flows. positive to the cathode-no current flows. The size of the current flowing depends on the potential difference between anode and cathode, but the valve does *not* obey **Ohm's Law-it conducts only in one direction.** If an alternating voltage is applied between anode and cathode instead of the high-tension battery, the current varies as shown in fig. 87.

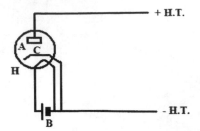

Fig. 86- The diode. A-anode,
C-cathode, H-heater, B-heater
(L.T.) battery

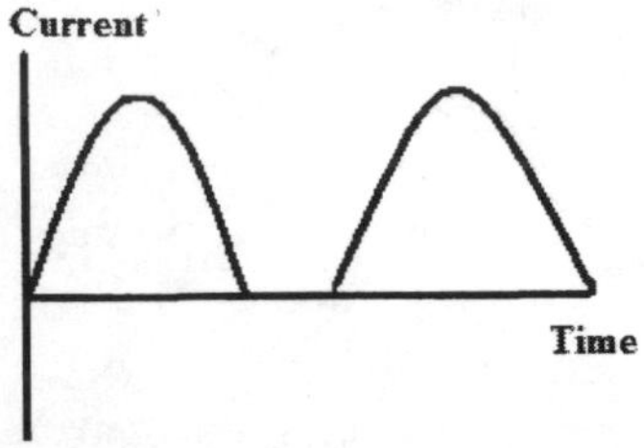

Fig. 87- Current-time curve
when H.T. battery in fig. 86
is replaced by a suitable
A.C. voltage